Hobart Pap

FARMING FOR FARMERS?

Farming for Farmers?
A critique of agricultural support policy

Richard W. Howarth
*University College of North Wales,
Bangor*

Published by
THE INSTITUTE OF ECONOMIC AFFAIRS
1985

First published in January 1985
by
The Institute of Economic Affairs
2 Lord North Street,
Westminster, London SW1P 3LB

© The Institute of Economic Affairs 1985

All rights reserved

ISSN 0309-1783

ISBN 0-255 36178-5

Printed in Great Britain by
GORON PRO-PRINT CO LTD,
CHURCHILL INDUSTRIAL ESTATE, LANCING, WEST SUSSEX
Set in 'Monotype' Bembo

Contents

		Page
PREFACE	Martin Wassell	xi
DEDICATION		xv
ACKNOWLEDGEMENTS		xv
THE AUTHOR		xvi

	Page
I. The 'Great Debate' on Agriculture	1
Change in the Political Climate	1
The Continuing Problem of the CAP	2
The Rise of the Environmentalists	3
The Spark	3
The Objectives	4
APPENDIX: *Monetary Compensatory Amounts*	5
II. The Growth of Agricultural Policy	6
19th-Century Foundations	6
The 1914-18 War	7
The Slump of the 1930s	8
France, Germany, and the USA	8
Britain and 1930s Protectionism	9
War and Post-war	10
Universal Support	13
III. The Nature of Agricultural Policy	14
Technical Policies	14
Research, Education and Advice	14

Support Policy	15
Types of Support	15
1. *Policies to raise and/or stabilise product prices*	16
2. *Policies to reduce the cost of inputs*	17
3. *Policies to support incomes directly*	17
4. *Policies of structural reform*	17
5. *International commodity agreements*	17

IV. The Economic Arguments for Support   19

The 'Special Case'	19
Special Features of Agricultural Production	19
Uncertainty	20
Instability	21
Low Income	24
Does It Matter?	28
Statistical and Presentational Difficulties	28
Total Income Understated	30
Standard of Living	32
Economic Well-being	33
The Demand for Food	35
Trend of consumption elasticities	36
Trend of growth in demand	38
Constraint on growth of farming industry	39
Income and substitution effects	41
The Farm Problem	43
The Rate of Exodus	47
Asset-fixity Theory	48
'Voluntary Acceptance of Negative Economic Rent'	50
Agriculture and the Balance of Payments	52

V. The Case Against Support   54

Support and the Farmer	54
Price Policy	54
Inequitable Distribution	61

Real and Relative Income Largely Unaffected 62
The Cost of Support 67
Distribution of the Burden of Support 75
The Benefits to Farmers 78
    Dissipation of support into land values 80
    500-acre millionaires 83
    Over-capitalisation 85
    Balloon theory 88
Economic Costs 89
Further Observations on the CAP 91
    Internal farm prices raised 92
    The scope for fraud 92
    Effects on the Third World 94
    Trade wars 96
    The agro-political dimension 99
    The destruction of the countryside 100

VI. The Politics of Agriculture 103
The General Climate of Opinion 103
The Agricultural Interest Groups 106
    Farmworkers and the NUAAW 106
    The National Farmers Union 107
    New allies 110
    The Ministry of Agriculture (MAFF) 111
The Agricultural Vote 113
    Voting significance 114
    Present political significance 116
Agriculture in Parliament 117

APPENDIX: List of Main Sources 120

VII. Future Policy 121
Summary of the Argument So Far 121
Food Security 122
Provisions Against 'Dumping' 124
Other Minor Functions 126

Agenda for Immediate Action	126
(1) Marketing boards	127
(2) Support for capital and other improvements	127
(3) Support for agriculture in special areas	128
(4) Public expenditure on research and development	129
(5) National price guarantees	129
(6) The Forestry Commission	129
Britain and the CAP: the Options	130
(1) To submit gracefully	130
(2) To press for 'reform' of the CAP	130
(3) To press for the scrapping of the CAP	131
(4) To withdraw from the CAP unilaterally	131
(5) To allow the CAP to die a natural death	131
The Decline of the CAP	131
Not Quotas	132
Differences in Philosophy	133
The CAP That Doesn't Fit	134
Towards a Free-Market Agriculture	137
Compensation scheme	138
The Pattern of an Unsupported Agriculture	140
FURTHER READING	143
SUMMARY	Back covers

LIST OF TABLES

I. (A) Agricultural Employment and Share of GDP Originating in Agriculture in Selected Countries, 1900–1962	26
(B) Relative Income Percentages in Selected Countries, 1900–1962	26
II. Role of Agriculture in National Economies and Relative Income Situation in 17 Countries, 1955–56 and 1963–64	27
III. Composition of Total Income of Couples and Individuals in the Trade Group Agriculture of the 1978–79 Survey of Personal Incomes	32

IV. Annual Food Consumption per Head in the
    United Kingdom, 1955-1977   37
V. Illustration of the Farm Problem in the Post-war
    Period (typical figures applicable to W. Europe
    and the USA)   45
VI. UK Indices of the Prices of all Agricultural Products,
    1935-1981 (1936-38 = 100)   58
VII. (A) Aggregate Net Farm Income and Farming
    Income in Money Terms and in Real Terms
    and per Full-time Farmer, 1937/38 – 1983   64
    (B) Real Farming Income per Full-time Farmer,
    1970-83   65
VIII. Relative Income of Agriculture: UK, 1955-56 – 1982   67
IX. Annual Percentage Variations in Real Income per
    Full-Time Farmer, 1959-60 to 1982-83   68
X. (A) Exchequer Support to Agriculture, 1954-55 – 1983   70
    (B) Exchequer Support to Agriculture at Current
    Values, at 1981 Values, and as a Percentage of
    Total Government Expenditure, '1956' – '1981'   71
XI. Some Estimates of UK Total Support Costs, 1956-1980   73
XII. Percentage Shares of Support Costs under Deficiency
    Payments System and CAP by Household
    Income Quartile Compared with Final Income
    Shares (1969)   76
XIII. Total Support Costs, Producer Gain and Farming
    Income, 1956-1980   79
XIV. Average Land Prices in England and Wales, 1943-1983   82
XV. Farm Rents as a Percentage of Net Farm Income,
    1952-1981   83
XVI. Comparative Income and Price Indices Affecting
    British Agriculture, 1964/5 – 1965/6 to 1982
    (1970 = 100)   86
XVII. The Number of MPs with Agricultural Affiliations
    Elected at the 1983 General Election   118

Preface

Mr Richard Howarth wrote this *Hobart Paperback* during the middle months of 1984 as Mr Arthur Scargill played out his epic battle to coerce the British taxpayer into an open-ended commitment to subsidise uneconomic mines until they were physically exhausted of coal. As the author observes, more than one commentator at the time drew an uncomfortable parallel (uncomfortable, that is, for the Government) between the vast sums of public money poured into Mr Scargill's bottomless pits and the even vaster subsidies devoured by Britain's agricultural industry. Certainly, some examination of farm support seemed timely.

Fourteen years after the IEA published his *Agricultural Support in Western Europe*,[1] Mr Howarth has returned to the subject for the Institute with a more comprehensive critique which incorporates the new dimension of Britain's participation for over a decade in the Common Agricultural Policy (CAP) of the European Community. His analysis and evaluation of British experience under the CAP make alarming reading not only for taxpayers, consumers, environmentalists and people concerned about the welfare of developing countries, but for a very large number of farmers as well. The more blatant sins and absurdities of the CAP – high food prices co-existing with butter mountains, wine lakes and other stockpiled surpluses – are regularly ridiculed in the British press and are better understood in Britain than on the Continent.

There are other shortcomings, however, which are less well-known and which Mr Howarth surveys with painstaking scholarship. For example, the CAP's effects are regressive for both consumers and farmers; it hurts the poorest consumers most and benefits the poorest farmers least. Secondly, the CAP damages the agriculture of many developing countries by 'dumping' European food surpluses on Third World markets and raising barriers around the Community to the Third World's food exports. And thirdly, the CAP's disruption of

---

[1] Research Monograph 25, 1971.

the agricultural industries of many efficient food-exporting countries – notably the USA – threatens to trigger a trade war which would reverse the gradual but highly successful liberalisation of international trade in manufacturing goods achieved since World War II.

The European Community is by no means alone in the world in wasting huge resources to protect its farmers. But the CAP is, nonetheless, a peculiarly distorting arrangement, and no economist would be surprised by its results. The mechanisms it employs – an open-ended commitment by official authorities to buy-in as much as their farmers care to produce if prices fall below a predetermined, artificially-high level; import levies to protect domestic production from cheaper foreign supplies; and 'restitution payments' (a euphemism for subsidies) to help the Community dump its otherwise unsaleable surpluses on third-country markets – are a sure recipe for very expensive trouble. It would be well-nigh impossible to sit down with pen and paper and design a system more calculated to stimulate production while at the same time curbing consumption. The CAP is a towering monument to the folly of distorting the price mechanism to achieve social policy objectives.

Fortunately, arrangements like the CAP contain the seeds of their own destruction in the longer term – though the process is admittedly costly. And there are plentiful signs at the present time that the CAP is destroying itself, signs which Richard Howarth explains and interprets as a 'withering-away' of the policy. As someone who applauds the wider achievements and objectives of the Common Market, he argues that Britain should do all in its power to hasten the withering-away as a preferable alternative to risking the break-up of the Community by a unilateral withdrawal from the CAP. Opinions will differ as to whether the CAP will wither away at all or fast enough following last year's agreement by the heads of the 10 governments to expand the Community's 'own resources'. What ought to be indisputable is that the uncontrollable profligacy of the CAP itself poses the biggest threat to the maintenance of the Common Market. Pro-Marketeers who look affectionately on the CAP as the first 'common policy' and major solid achievement of the Community should reflect that it is nowadays the source of endless political wrangling among the member states which not only sours the prospects for collective progress on other fronts but pre-empts much of the time of senior ministers and officials for such visionary matters as the allocation of dairy quotas and the pesticide content of sheep carcases. It is becoming increasingly probable that, if the CAP is not scrapped, it will tear the Community apart.

Although much of Mr Howarth's *Paperback* is inevitably focussed on the CAP – since British policy is essentially subsumed in it – his principal thesis is that large-scale agricultural 'support' by whatever means (though some means are less wasteful and distorting than others) aggravates '*the* farm problem' of an over-supply of farmers while failing to raise the average incomes of farmers relative to incomes outside agriculture. He examines the multitude of arguments advanced over the decades which purport to show that agriculture is a special case worthy of large-scale subsidisation – and finds them all specious or defective. Moreover, as he strikingly demonstrates, the effects of support on farmers themselves have been extremely haphazard and inequitable. Most of it has been dissipated in higher land values and rents. The chief net beneficiaries have been the larger and better-off farmers, and especially long-time landowners. Tenant farmers and new entrants to the industry have lost out.

If agricultural support policy burdens the taxpayer, harms the consumer (or at least does him no good), and benefits only a limited number of farmers while injuring others, why have one? Having posed this question, the author proceeds to outline a programme for a phased withdrawal of generalised support, accompanied by selective and temporary compensation payments for members of the farming community who would suffer hardship. Government would retain some functions: the setting and enforcement of standards for public health and hygiene and animal health and welfare; control of the use of chemicals and drugs; and the collection of statistics. Government would also have to ensure that the country's minimum food requirements were secured for strategic reasons and that other countries were not permitted to 'dump' their food surpluses on the British market. Apart from these, however, at the end of the withdrawal period agricultural prices, output and incomes would be determined by competition in the market.

Will Richard Howarth's analysis and proposals galvanise the politicians into action? He anticipates that question with a fascinating section on the 'politics of agriculture'. The political influence of farmers is indisputable, but he shows that it no longer derives from voting strength. Rather it survives because of incomparably well-organised lobbying and powerful friends of farming in ancillary industries, Parliament and, indeed, the Cabinet.

All that remains to say is that, whatever happens, the current 'Great Debate' on agriculture is unlikely to fizzle out so long as controlling the bulldozer of public expenditure remains a major headache for governments – and an analgesic for that malady is nowhere near in sight. The intellectually strong case for dismantling the agricultural support system presents an acutely uncomfortable challenge to the governing political party of this country today, with its traditional links with the farming community and its belief in competitive private enterprise and limited government. But, as Richard Howarth says (and to return to Mr Scargill):

> 'It is hazardous for people in glass houses to throw stones. Politicians who wish to reform trade unions and curtail their restrictive practices, and who seek to reduce subsidies to nationalised monopolies and return many of them to private ownership, must at the same time be willing to attack the monopolies, restrictive practices and subsidies enjoyed by their supporters – be they opticians, solicitors, barristers, accountants, architects or, above all, farmers. It is in this sense that farm support is the biggest hindrance to economic progress'.

The constitution of the Institute obliges it to dissociate its Trustees, Directors and Advisers from the analysis and conclusions of this *Hobart Paperback*. However, it offers the study as a comprehensive and thought-provoking contribution by a noted academic authority to public discussion of an enormously costly and highly topical area of public policy.

January 1985　　　　　　　　　　　　　　　　　　　MARTIN WASSELL

Dedication

To Jenny for her patience with me during the long, hot summer of 1984 when most of this paper was written; to Gwen Birkett for her painstaking work in typing and checking successive drafts of the manuscript; and to the agricultural economics students at Bangor who have over the years acted as a sounding-board against which to test my views.

Acknowledgements

I owe a particular debt of gratitude to Mr Paul Cheshire of the Department of Economics, University of Reading, and to Mr George Peters, Director of the University of Oxford Institute of Agricultural Economics, for their encouragement, advice, and constructive comments on an early draft of this book.

R.W.H.

The Author

RICHARD HOWARTH was born in Huddersfield, Yorkshire, in 1940 and spent his formative years on his parents' dairy farm near York. He was educated at Giggleswick School and St Andrews University from which he graduated in 1963 in Moral Philosophy and Political Economy. He went on to study agricultural economics at the Oxford University Institute of Agricultural Economics from which he gained the Diploma in Agricultural Economics in 1964. Since then he has lectured in Agricultural Economics at the University College of North Wales, Bangor. His main interests are in national and international agricultural policy and the politics of agriculture.

He has had articles published in *Political Studies*, the *National Westminster Bank Review*, and *Economic Affairs*. For the IEA he contributed to *Essays in the Theory and Practice of Pricing* (1967), and wrote *Agricultural Support in Western Europe*, Research Monograph 25 (1971). He was a member of the Agriculture Working Party of the Omega Project sponsored by the Adam Smith Institute in 1982-83 and wrote the sections on the CAP of its report on agricultural policy. He has been a frequent contributor to the farming press and has acted as an adviser to the Consumers' Association, national newspapers, and television companies.

I. The 'Great Debate' on Agriculture

In modern times, agricultural policy has been in continuous existence for at least 55 years in most developed countries, and in many for far longer. Taxpayers' expenditure on it had by 1982 reached around US $14 billion a year in both the USA and the European Community (EC). As these milestones have been reached, the despair of critical students of this phenomenon that it would ever attract serious public debate has at last been extinguished – and with a vengeance.

The debate in Britain began in late 1982, gathered pace rapidly in 1983, and showed no sign of losing momentum in 1984. The forum is a large one consisting of a continuous spate of articles and correspondence in the serious national press; a few articles in the popular press; four major television documentaries in the past year or so; a number of reports from independent research institutes; several new books examining farm support from different angles; commentaries on them by academics; and, of course, a flood of defensive material from the agricultural lobby.[1]

Two questions immediately arise: 'Why did it take so long for this "Great Debate" to begin?' and 'What sparked it off?' The answers to the first question will be provided in later sections; but the long period of public consensus (or, really, public ignorance and apathy) about farm policy was basically a consequence of the effectiveness of the agricultural lobby, the ineffectiveness of opposition to it, and the long-running efforts of politicians of all parties to woo the farmers' vote by vying with each other in their generosity to farmers.

Change in the Political Climate

The spark to the debate required a combination of events and favourable circumstances. Of the latter, the first was the change in the broad political climate with the advent of Mrs Thatcher's Government in 1979. She was pledged to the principles of the market economy,

---

[1] The author has collected over 200 relevant press cuttings during the past 18 months and would not claim that this is exhaustive. *World in Action*, the BBC's *Panorama*, Thames TV's *Against the Grain*, and Anglia's *The Food War* are the television programmes referred to. The various reports and publications are identified later in the text.

cutting public expenditure, eliminating waste, and encouraging individuals and businesses to stand on their own feet. Those who had previously felt it was futile to criticise the farm support system now believed they had a better chance of a sympathetic hearing and government action. In the event they were, at first, disappointed. The Government had more important nettles to grasp – particularly inflation. Far from reducing farm support, it acquiesced in the excesses of the Common Agricultural Policy (CAP) and, indeed, increased them by maintaining a positive British Monetary Compensatory Amount (MCA)[1] which, because it kept farm product prices in Britain above those in most other EC countries, was effectively a further tax on food. There are signs, however, that the second Thatcher Government is taking a much harder look at such sacred cows as agriculture, following its recognition of the 'public expenditure time-bomb'.[2]

The Continuing Problem of the CAP

The second circumstance is the CAP itself. The European Community's first major policy, which long ago became its major burden and source of dissent, has attracted attention not simply because of its ever-increasing budgetary cost but because of its additional burdens on consumers, its conspicuous waste, its sheer inequity and its potentially disastrous external consequences for world trade. Without the

---

[1] Discussed in the short Appendix to this Section, p. 5.
[2] 'The public expenditure time-bomb' is a phrase used in the press and by Mrs Thatcher. It refers to the fact that welfare spending of various types in Britain (and also in France, Holland, Sweden, Italy and West Germany) is rising rapidly and is now 53 per cent (compared with 43 per cent 10 years ago) of total government expenditure – which itself has risen between 1979 and 1983 from 40·5 to 43.5 per cent of GNP. NHS spending alone has risen from £7·5 billion in 1978-79 to £15·5 billion (estimated) in 1984, whilst NHS manpower has increased by 200,000 over the past 10 years. As health technology improves, demands on the NHS increase inexorably – apart from the demands placed on it by a growing number of OAPs who are expected to increase from 17·9 per cent of the total population in 1981 to 22·4 per cent in 2031, not allowing for a reduction in the retirement age. The implications for the cost of pensions are also frightening, with the advent of earnings-related pensions adding £2 billion to the pension bill in 1998 and the possibility that National Insurance contributions will increase by 35 per cent between 2011 and 2033 when the number of workers supporting each pensioner will fall to 2·6 in 2031 from 3·4 in 1981. As a result, new thinking on welfare funding is required. Furthermore, non-welfare government spending will be squeezed and will have to be more closely scrutinised than ever before. This squeeze has been apparent for a long time; over the past 10 years, for example, defence spending has fallen from 15·1 to 13·2 per cent of total government spending, and other expenditure (including agriculture) has fallen from 41·9 to 33·6 per cent.

CAP, the whole long saga of Britain's EC budget contribution and the attendant aggravation would not have existed.

The Rise of the Environmentalists

The third factor has been the rise in the 1970s and 1980s of the environmentalists, ecologists, and 'Greens' from the ranks of cranks and oddities to become a serious political force in some countries and an effective pressure group in others. This development has added a new dimension to the criticisms of farm support which were previously mainly economic. The environmentalist critics have fired an impressive broadside at the farming community from which it is still reeling. From being the guardian of the countryside, the farmer is alleged to have become its destroyer. The environmentalists believe that farm policy has encouraged him in this destruction. The recent attack has been led by polemicists such as Marian Shoard[1] and Richard Norton-Taylor,[2] and pursued by more academic writers such as Bowers and Cheshire.[3] Lord Melchett, in his presidential address to the Ramblers' Association in the spring of 1984, aimed another powerful blast at the farmers for desecrating the countryside.

The Spark

Against this background it required only the right spark to kindle the potential fire of a far-reaching public debate. The spark was generated in December 1982 by Richard Body's book *Agriculture: The Triumph and the Shame*,[4] which was remarkable as much for its author as for its content. Richard Body has all the attributes which would be expected of the staunchest member of the agricultural lobby: a Tory, an MP since 1966 for one of Britain's most agricultural constituencies (Holland with Boston), a former farmer, stock breeder, and National Farmers Union (NFU) member. Yet in his book he revises all his former views and produces a devastating polemic attacking the whole concept of support from both the national and the farmers' point of

---

[1] M. Shoard, *The Theft of the Countryside*, Temple Smith, London, 1980.
[2] R. Norton-Taylor, *Whose Land is it Anyway?*, Turnstone Press, Wellingborough, 1982.
[3] J. K. Bowers and P. Cheshire, *Agriculture, the Countryside and Land Use*, Methuen, London, 1983.
[4] R. Body, *Agriculture: The Triumph and the Shame*, Temple Smith, London, 1982.

view. Although many of the views he expresses are not original,[1] the timing and the source caused an explosion in British agricultural circles the reverberations of which are still being felt.

Sadly for a professional agricultural economist, the debate was not started by a member of the profession even though much of the requisite ammunition has long been available in its specialist publications. As Mr George Peters (Director of the Oxford Institute of Agricultural Economics) acknowledges in his review of Body's book in the *Journal of Agricultural Economics*,[2] British agricultural economists have singularly failed to make an impact on public awareness of agricultural policy issues. Some of the reasons for their reticence are mentioned later, but Body's book has encouraged several of us to step closer to the limelight and make a public response to his views.[3] *The Triumph and the Shame* has also undoubtedly drawn attention to other relevant publications, particularly the work of the Institute of Fiscal Studies,[4] Bowers and Cheshire,[5] and the Adam Smith Institute.[6]

The Objectives

The objectives of this *Hobart Paperback* are to advance the debate on agriculture by re-examining the nature and scope of support policy and the arguments used to justify it; to question those arguments by emphasising facets which have been underplayed or neglected, paying particular attention to the CAP; to discuss the true reason for government intervention, namely, the politics of agriculture; and to consider the future of agricultural policy.

---

[1] Certain academics have for some time been saying similar things to those said by Richard Body; for example, John Weller: 'Agriculture is two-faced: one face projects an image of efficiency second to none, while the other of an industry, like the social services, requiring public support in order to exist'. (J. Weller, *Modern Agriculture and Rural Planning*, The Architectural Press, London, 1967, p. 202.)

[2] G. H. Peters, Review of *Agriculture: The Triumph and the Shame* in *Journal of Agricultural Economics*, Vol. xxxiv, No. 3, September 1983.

[3] A seminar on Body's book was held at the Royal Society on 14 June 1983. Its proceedings are published in *Agriculture: The Triumph and the Shame: an Independent Assessment*, CEAS/CAS, Wye College and Reading University, June 1983.

[4] C. N. Morris, 'The Common Agricultural Policy', *Fiscal Studies*, March 1980; A. W. Dilnot and C. N. Morris, 'The Distributional Effects of the Common Agricultural Policy,' *Fiscal Studies*, July 1982.

[5] *Op. cit.*

[6] *Omega Report: Agriculture Policy*, Adam Smith Institute, London, 1983.

APPENDIX TO SECTION I

Monetary Compensatory Amounts

Monetary Compensatory Amounts (MCAs) are one of the mechanisms of the CAP and were introduced in the late 1960s to protect farmers from currency fluctuations. Without them, farm product prices in an EC member country would rise if the country's currency fell, and would fall if the currency rose, because EC farm prices are fixed in terms of a basket of all EC currencies (the European Currency Unit (ECU)). MCAs ensure that currency movements are not reflected in national farm prices unless a national government wants them to be by 'devaluing' or 'revaluing' its 'green' currency. Many governments, including the British and the German, have over long periods refused to keep the green rate in line with the market rate. Strong-currency countries, like Germany, have continuously had food prices above the EC average; weak-currency countries, like Italy, below the average. Britain maintained an over-valued green rate during the mid-1970s when the pound was falling, and therefore below-average food prices. Since 1979, when the pound has been stronger, she has maintained an under-valued green rate and therefore above-average food prices. MCAs bridge the difference between national price levels and thereby maintain a common EC trading price in farm products. They can act as either taxes or subsidies. They subsidise exports from high-price to low-price member countries, but tax exports in the opposite direction. MCAs are an item in the budget of the EC agricultural fund (EAGGF or FEOGA)[1] and have implications for member states' budgetary contributions.

The Economist, 5 April 1980, p. 49, from which this note has been abstracted, provides a useful brief discussion of MCAs. A more detailed treatment is given in A. Swinbank, *The British Interest and the Green Pound*, CAS Paper 6, University of Reading, June 1978.

---

[1] EAGGF is the acronym for the European Agricultural Guidance and Guarantee Fund; FEOGA is the French acronym for the same fund.

II. The Growth of Agricultural Policy

Nineteenth-century Foundations

Agricultural policy, in its broadest sense, refers to the actions of government in the sphere of agriculture. In that sense, agricultural policy has a very long history going back at least to the ancient Egyptians, Greeks and Romans. In England, the 'Corn Laws' were in existence from the early Middle Ages to regulate internal trade and to restrict exports of grain. From 1660 they also regulated imported grain by means of specific tariffs, until their celebrated repeal in 1846 by Sir Robert Peel. France and Germany, too, have had a long history of protection of grain.[1]

The 19th century ushered in huge technological improvements in agriculture; drainage, livestock breeding, agricultural chemistry, and mechanisation all made major strides. The years from the mid-1840s to the mid-1870s are known as the 'Golden Age' of British agriculture – the period of 'high farming'.[2] Whilst improvements were widely adopted in Britain, agriculture on the Continent had progressed little from feudal times to the mid-19th century. There was little international agricultural trade except within Western Europe where it was still very small and mainly in grains. Trade in livestock products had barely commenced. Russia was still the world's biggest grain exporter. The vast potential of American agriculture was still to be released.

Just before the period characterised as the Great Depression (1880-1900), there was an interlude of free trade – led by Britain under the influence of the Manchester School of Free Trade – which spread throughout Western Europe. It went against long-standing traditions of protectionism in France, Germany, and Italy; and it was short-lived. From the 1870s the Americans, with the Civil War behind them, began rapidly to increase their grain exports to Western Europe (from an

---

[1] The history of agricultural policy is told admirably by Michael Tracy in his book *Agriculture in Western Europe*, Granada, London, 2nd edn. 1982, from which the material for much of this Section was gathered.
[2] Lord Ernle, *English Farming Past and Present*, Heinemann, London, 6th edn. 1961, Chap. xvii.

annual average of 5 million bushels in 1851-60 to 184 million bushels in 1895-99). Shipments of frozen meat from the USA commenced in 1875, and from Australia in 1877. From then onwards, Britain took advantage of this 'cheap' (that is, efficiently produced) food. Most of the Continent, where agriculture was still politically dominant, was totally unable to compete and therefore resorted to a revival of protectionism, which swept through France, Germany, Italy and Belgium in the late 1870s and the 1880s, and continued to the First World War. According to Tracy,[1] 'the degree of protection [in 1913] lay somewhere between 20 per cent and 30 per cent'. It was on these late-19th-century foundations that the edifice of the Common Agricultural Policy was built in the second half of the 20th century.

The 1914-18 War

British agriculture was forced to face the rigours of *laissez-faire* during the Great Depression, went through a traumatic adaptation, and emerged leaner and fitter having shed labour rapidly and accelerated the trend away from arable towards livestock production.[2] Denmark and Holland also held to their free-trade principles and became the two most efficient and competitive exporters of livestock products in Western Europe. Despite the campaigns for 'Fair Trade', 'Tariff Reform', and 'Imperial Preference', both major political parties in Britain (Conservative and Liberal) remained officially committed to free trade until (and after) the First World War. Indeed, Joseph Chamberlain's abortive tariff campaign cost the Conservatives the 1906 General Election which was won by the Liberals on a platform of 'no taxes on food'.

The origins of recent agricultural policy in Britain lie in the 1914-18 War. Up until the early years of the Great War, the country had total confidence in the ability of the Royal Navy to protect the two-thirds of its food supply which was imported by sea, mainly from the

---

[1] M. Tracy, *op. cit.*, p. 25.
[2] Grain prices were halved and livestock prices fell by a third during the Depression. The proportion of the occupied population of Britain engaged in agriculture fell from 18 per cent in 1861 to 8 per cent in 1911. The proportion of gross agricultural output comprised by livestock products rose during the Depression from 55 to 70 per cent whilst the arable product proportion fell from 45 to 30 per cent. British agriculture thus concentrated on those products in which it had a comparative advantage and where demand would grow most as affluence increased.

Empire. That belief was to be shattered by a new machine of war. If any one event can be said to be the origin of British agricultural policy in recent times it was the introduction of the submarine.

Having entered the war without plans for food production, the threat of shortages forced the then government, after considerable debate and delay, to encourage the production of wheat, oats, and potatoes by means of guaranteed prices, which were introduced in February 1917. For the first time since 1846, the policy of *laissez-faire* was overthrown. After the war there was disagreement about continuing the guarantees, which the farmers were naturally very keen to retain. Whilst some people genuinely believed we were dangerously dependent[1] on imported food and that farmers' incomes should be supported to encourage a certain level of production, others (the majority) wanted to return to *laissez-faire*. A Royal Commission, reporting in 1919, was split 12 to 11 in favour of continuing guarantees. Bowing to farmers' demands, the government introduced the Agriculture Act of 1920 to continue the guarantees indefinitely – which proved to be a very short time indeed since they were ended on repeal the following year after their cost had risen to what was then considered an exorbitant £18 million. The 'Great Betrayal' of 1921 coloured the attitude of farmers to governments for many decades thereafter.

The Slump of the 1930s

Britain returned to a free market in agriculture for the remainder of the 1920s (except for sugar-beet), whilst France, Germany and Italy all re-introduced protection for their farmers by tariffs and other restrictive devices. But it was the Wall Street crash of 1929 and the ensuing slump of the 1930s which brought about the wholesale growth of interventionism and the development of new methods of supporting agriculture.

France, Germany and the USA

France backed up tariffs by quotas. Draconian restrictions were imposed on vineyards and a panoply of additional supports for

---

[1] T. H. Middleton, *Food Production in War*, Oxford, 1923. Middleton points out that in 1918 self-sufficiency in food was still equivalent to only 155 days in the year, compared with 125 in 1914. Also Tracy, *op. cit.*, for further references to the strategic argument put forward at that time.

wheat, including a milling ratio, denaturing subsidies,[1] acreage quotas, and export subsidies. From 1933, Hitler's Germany set up the most complete and complex system of government control of agriculture (with the exception of the present Communist countries) covering every detail from prices, production, imports and exports to the control of every aspect of the life of the farmer and his family. Agriculture was to be supported, supervised and cosseted not only to ensure self-sufficiency in food in war-time but to provide the breeding stock of the pure Teutonic race.[2]

American agricultural policy, which has changed little since those days, also began in the slump with the establishment in 1929 of the Federal Farm Board and in 1933 (under the Agricultural Adjustment Act) of its successor, the Commodity Credit Corporation. The Commodity Credit Corporation had powers to operate a support buying system to maintain farm product prices. It thus became the first government agency to accumulate costly food mountains, which it has been doing on and off ever since, and from whose example those responsible today for the CAP could learn a lot but learn nothing. It was also one of the first agencies to attempt to control supply with equally unsuccessful results, and which the CAP is blindly emulating with its recently agreed milk policy.

Britain and 1930s Protectionism

The abandonment of *laissez-faire* in Britain was part of the general stampede into protectionism of the 1930s; intervention in agriculture was largely a part of the general assistance for all industries. But tariffs could not be introduced for agricultural products as they were for industrial products because the principle of Imperial Preference, embodied in the Ottawa Agreements of 1932, exempted from duty the bulk of our food imports which came from Empire sources. Other methods of assisting agriculture had to be devised hurriedly, product by product, in a time of mounting crisis. As it turned out, these piecemeal solutions of the early 1930s formed the basis of British policy not only for the remainder of the decade but for the whole of

---

[1] 'Denaturing' means rendering unfit for human consumption. The product is then sold off cheaply for animal feed or other purposes.

[2] For an interesting description of European agriculture in the 1930s, P. L. Yates, *Food Production in Western Europe*, Longmans, Green, London, 1940.

the post-war period until we began to adopt the EC régimes in 1973.

The main mechanisms adopted included guaranteed prices by deficiency payments[1] for wheat adjusted in relation to a 'standard quantity' or quota (The Wheat Act 1932); quantitative restrictions on imported bacon and beef; import duties on beef and veal; tariff and trade agreements with Denmark, Argentina and the USA; a flat-rate subsidy on cattle (1934); and the introduction of producer-controlled marketing boards, under the Agricultural Marketing Acts of 1931 and 1933, for milk, pigs, potatoes and hops.

Yet, despite the array of legislation and support measures, the total effects were relatively slight. There was a major shift in imports from 'foreign' to Empire sources, but the volume of food imports rose by 6 per cent between 1927-29 and 1938. Farmers' prices and incomes remained depressed, recovering only as the general recovery took place. And the costs to the taxpayer remained surprisingly small – estimated at £16 million in 1935-36 and £11·5 million in 1937-38.[2]

War and Post-war

It was undoubtedly during the Second World War and its aftermath that agricultural policy as we now know it became deeply entrenched. The experience of the First World War had made us aware of the vulnerability to submarines of our overseas supplies. We entered the Second World War with comprehensive plans for agricultural expansion through guaranteed prices for all agricultural products, rationing and price control of consumer purchases, and strategic stockpiles of food. Our population survived the war considerably better-fed than our Continental enemies or allies, many of whom were in a parlous condition by 1945. Agriculture in Britain concentrated on the staple foods (cereals, potatoes, milk) and undertook a 50 per cent expansion of the arable acreage. Our war-time agriculture was a success story for both consumers and farmers.[3] It was at this time that a close partnership was forged between farmers (the National Farmers

---

[1] Below, page 11.
[2] Source: E. H. Whetham, *British Farming 1939-49*, Nelson, London, 1952. Despite its title, this book gives a good history of agriculture from 1919 to 1949, including the crucial war years and the post-war legislation.
[3] R. J. Hammond, *Food: The Growth of Policy* [*History of the Second World War*], UK Civil Series, HMSO, London, 1951.

Union) and government (the Ministry of Agriculture) which was to survive long into the post-war period.

When the war ended, the military threat was replaced by an economic threat. There was anxiety particularly about the balance of payments, and international food continued for several years to be in short supply and expensive. In these circumstances, the NFU was in a powerful position to obtain continued support for agriculture from a grateful nation. Politicians of all parties differed only in the extent of the generosity they would bestow. In the event, it was a Labour Government which laid down British policy in the Agriculture Act of 1947 and its associated Expansion Programme (1947-52). This Act was a comprehensive piece of legislation whose detailed objectives and provisions may be found in all the textbooks. The vague objectives (stability, efficiency, 'producing such part of the nation's food as in the national interest it is desirable to produce in the United Kingdom', 'proper remuneration for farmers', and an 'adequate return on capital') were backed up by specific measures to achieve them: an Annual Review of the industry, to be carried out jointly by government and farmers, in the light of which guaranteed prices were to be fixed for 11 major products whose markets also were to be 'assured'.

Until 1952-54 the guaranteed prices and assured markets were effected, as during the war, by the government purchasing through the Ministry of Food all that farmers could produce. In many cases the products were sold into the distribution system at lower prices, and these consumer food subsidies amounted to £465 million in 1949. The 1951 Conservative Government, which was pledged to abolishing both food subsidies and rationing, gradually freed the consumer markets in agricultural products, but maintained the commitment to guaranteed producer prices by introducing an extremely complex new system, based on pre-war mechanisms. Briefly, cereals and fatstock were to have a deficiency payments system under which farmers sold their output for the best market price they could obtain and received retrospective payment for each unit of output from the state to make up any shortfall between the average market price and the guaranteed price determined at the Annual Review. Such schemes covered some 40 per cent of total output. New marketing boards and others revived from pre-war times, most of which had statutory monopoly powers to manipulate supplies and prices, implemented the guarantees for milk, wool, hops, potatoes, and eggs (from 1957). The boards covered 44 per cent of the output by 1957-58. The small remaining output was subject to free markets (poultry meat, for example), import duties (horticultural products), or a special scheme (sugar beet).

By the mid-1950s, some of the gluts of the 1930s began to re-appear for eggs, pigs, and milk. The NFU became extremely worried that the rising cost of support might lead to another 'Great Betrayal' after conflict over the 1956 Review. In 1957 it embarked on a campaign to win 'long-term assurances' for agriculture which led to what was possibly the last great triumph of NFU strength, the Agriculture Act of 1957. Under this Act the Conservative Government effectively hamstrung itself and its successors (unless they dared repeal the Act) in their legal ability to alter agricultural prices and other supports, by undertaking (a) not to reduce the total value of the guarantees by more than 2·5 per cent in any one year, after allowing for any changes in production costs, and (b) not to cut the guaranteed price of any product by more than 4 per cent in any one year, or for livestock products by more than 9 per cent in any three-year period. Together, the 1947 and 1957 Acts led to mounting support costs[1] and considerable embarrassment for successive governments in the 1960s and early 1970s prior to Britain's entry into the EC. One element of the so-called 'New Policies' to limit the exchequer cost of the early 1960s – the last major policy change before EC entry in 1973 – was to extend the 'standard quantity' system to most products whose prices were guaranteed, thereby limiting the full guaranteed price to a government-fixed volume of output and reducing the price in proportion to 'excess' output. This system was another forerunner, highly unpopular with farmers, of the present attempts to control supply in the EC. The other element of the 'New Policies' was the minimum import price régime for cereals, backed up by residual import levies – an incredible set of 'arrangements' with overseas suppliers under which they kindly agreed to raise the prices of the cereals they were offering us from

---

[1] The total exchequer cost of support rose from £197·3 million in 1954-55 after de-control, to £288 million in 1957-58; it fell back to £263·4 million in 1960-61 before rising steeply to £343·2 million in 1961-62; it then fell back to around £300 million in the late 1960s and early 1970s, rising to £392 million in 1973-74, the year of Britain's entry into the EEC. Sources: *Annual Review and Determination of Guarantees 1969*, Cmnd. 3965, HMSO, London, March 1969; and *Annual Review of Agriculture 1975*, Cmnd. 5977, HMSO, London, March 1975.

around £17 to £20 per ton. Thus, to avoid the embarrassment of having to ask Parliament for more money, Ministers, both Conservative and, later, Labour, were proud to raise the price of food and feedingstuffs at home and to transfer money across the exchanges into the pockets of our foreign competitors whilst at the same time talking of a balance-of-payments crisis. Only in the name of agricultural support could such economic eccentricities be perpetrated.

Universal Support

Governments in the USA, France and Germany also gave their farmers price and income guarantees during the post-war period. Indeed, OECD reports describe various methods of government assistance in all its member countries.[1] In nearly every instance, both the complexity and amount of support have steadily increased over the years. This alarming and rising trend was given a further boost by the advent in 1962 of the CAP.

The present problems of agriculture have long been foreseeable and foreseen. They are the culmination of a trend which has long worried observers of international trade who have noted the marked contrast between the post-war liberalisation of world trade in industrial products under successive rounds of the General Agreement on Tariffs and Trade (GATT) and the situation with agricultural products where, as the late Professor Harry Johnson put it:

'... the protection of domestic producers by tariffs and especially by import quotas, subsidies, and other governmental assistance, is rampant and has been perceptibly increasing over the post-war period'.[2]

Agricultural protection has now become a major cause of the present threat to *industrial* free trade.

---

[1] Details of the reports are given in footnote 1, page 16.
[2] H. G. Johnson, 'On Demolishing the Barriers to Trade', in *Rebuilding the Liberal Order*, Occasional Paper 27, IEA, London, 1969, p. 15.

III. The Nature of Agricultural Policy

Originating in depressions and wars as part of a general move to protectionism or to safeguard the food supply, agricultural policy has developed and expanded in the post-war period to a degree which would have been unimaginable in the pre-war period. The variety of subjects which can be included in its ambit fall into three broad categories: technical policies; research, education, and advisory services; and support policies.

Technical Policies

Technical policies cover, amongst others: farm safety; animal health and welfare regulations; hygiene rules; controls on the use of farm chemicals, medicaments, and drugs; measures for controlling pests and vermin; and legislation on minimum wages and working conditions for farm workers. All these are similar to regulations covering the operations of other industries (the Factories Acts, and so on) which have also mushroomed in the post-war period. Most are relatively uncontroversial and have either low or hidden costs. Some of them, such as animal welfare and minimum wages, do arouse debate outside the agricultural industry from time to time but, on the whole, they are matters left to the industry to sort out.

Research, Education and Advice

Research, education, and advisory services are important activities in which governments are concerned and which do have considerable costs. Research on agriculture and food commissioned by the MAFF and its agencies, for example, is expected to cost £58·9 million in 1983-84.[1] Again, research and education policies do not differentiate agriculture from other spheres and are relatively innocuous. This is not to deny that both are in need of close scrutiny; merely that here is not the place to do it. Having its own free state-provided advisory

---

[1] HM Treasury, *The Government's Expenditure Plans 1984-85 to 1986-87*, Vol. II, Cmnd. 9143-II, 1984, p. 23.

service (the Agricultural Development and Advisory Service (ADAS)), costing £61 million in staff costs in 1982-83, does distinguish agriculture from most other industries, although such services are common for agriculture in other countries. Whatever the justification for free advice in war-time, there cannot be a case for it today. If the state wishes to compete with private advisers it should do so on a self-financing basis.

Support Policy

Support policy, sometimes misleadingly referred to as 'the farm subsidies', is the most far-reaching aspect of government involvement in agriculture and the one which arouses political controversy because of the burdens and benefits involved. A stipulative definition of agricultural support policies is 'government policies designed to raise and/or stabilise *per capita* farm incomes'. Farmers' incomes, rather than other general national goals, have become the objective in the post-war period: raising *per capita* farm incomes above their free-market level and evening-out their flow over time more than without intervention. The term '*per capita*' is used deliberately because not all policies are designed to raise *total* farm income. Thus structural policies aim to increase the incomes of a smaller number of farmers by sharing out a given size of cake in larger slices among a smaller number of recipients.

The main characteristic of a support policy, therefore, is that it is financially discriminatory in favour of agriculture and that the finance comes from the rest of the population as either taxpayers or consumers. This assistance benefits agriculture in either or both of two ways: first, by favouring agriculture in competition with other economic activities in individual national economies; and, secondly, by favouring the agriculture of a particular country (or group of countries, as with the CAP) in competition with the agricultures of other countries. An inevitable result of such discrimination is that support becomes an object of political controversy within and between nations.

Types of Support

Agricultural support has been implemented by numerous, ingenious, and complex methods. Indeed, it would be possible to write a full-

length book simply cataloguing support measures.[1] Basically, however, the methods relate to the determinants of *per capita* farm income:[2]
(a) Farm product prices;
(b) The quantity of farm production;
(c) The prices of agricultural inputs;
(d) The quantity of inputs to produce a particular volume of output;
(e) The number of farmers.

Mathematically, if y is the aggregate income of the agricultural sector, then:

$$y = [(a) \times (b)] - [(c) \times (d)]$$

and *per capita* income $= \left[\dfrac{(y)}{(e)}\right]$

Policy-makers act upon any of the above variables or combinations of them in their efforts to alter farm incomes; they can also directly subsidise income itself rather than trying to affect it indirectly *via* its determinants, although this is rarely done. To give some idea of the scope of agricultural support, it is worth repeating the categorisation of support measures given in a previous IEA publication:[3]

1. *Policies to raise and/or stabilise product prices*
    (i) Increasing the demand for farm products (e.g. advertising);
    (ii) restricting the entry of imported produce (e.g. tariffs, quotas, variable levies);
    (iii) restricting domestic supplies (e.g. intervention buying, production and marketing controls);
    (iv) operating buffer stocks;
    (v) subsidising exports;
    (vi) price discrimination (e.g. milk marketing in the UK);
    (vii) subsidies on product prices (e.g. deficiency payments);

---

[1] The OECD has published a number of book-sized reports describing agricultural policies in member countries. Examples are: OEEC (forerunner of the OECD), *Agricultural Policies in Europe and N. America: Price and Income Policies*, Paris, 1957. OECD: *Trends in Agricultural Policies since 1955*, Paris, 1961; *Low Incomes in Agriculture: Problems and Policies*, Paris, 1964; *Review of Agricultural Policies, 1977*, Paris, 1978.

[2] There is a useful discussion of these variables in J. Marsh and C. Ritson, *Agricultural Policy and the Common Market*, Chatham House, PEP, European Series No. 16, London, 1971.

[3] R. W. Howarth, *Agricultural Support in Western Europe*, Research Monograph 25, IEA, London, 1971.

(viii) providing market intelligence;
(ix) provisions to encourage farmers' organisations (e.g. co-operatives, marketing boards) which may carry out one or more of the above, or to enhance farmers' bargaining power.

2. *Policies to reduce the cost of inputs*
   (i) directly subsidising the inputs of agriculture (e.g. capital grants, fertiliser subsidies);
   (ii) refraining from taxing agricultural inputs (e.g. agricultural land) or taxing at preferential rates (e.g. fuel oil);
   (iii) providing credit at preferential interest rates;
   (iv) provisions to encourage groups or co-operatives for joint purchase of supplies or credit.

3. *Policies to support incomes directly*
   (i) cash grants to farmers according to farm size (small) or location (upland/marginal areas);
   (ii) special subsidies on the product or inputs of small/marginal holdings (e.g. EC less-favoured area (LFA) hill cow and hill sheep subsidies).

4. *Policies of structural reform*
   (i) consolidation of 'fragmented' holdings;
   (ii) amalgamation of small holdings;
   (iii) pensioning-off older farmers;
   (iv) paying re-training and re-location expenses for people leaving the land;
   (v) land reform (mainly in developing countries).

5. *International commodity agreements*
   (i) export-restriction agreements;
   (ii) international buffer stocks;
   (iii) multilateral contract agreements.

Some idea of the incredible complexity which farm support has reached can be given by indicating that the CAP involves all of the first 19 categories listed above with many permutations on most of them spread across the wide variety of products it covers. The overwhelming emphasis, however, remains on farm product prices since 'end-prices' are what farmers are most interested in, with subsidised inputs taking second place.

Section V examines the overt and hidden costs of this panoply of support. Suffice it to say here that the CAP now has an annual budgetary cost of some £10 billion and that the total budgeted programme of the MAFF for 1983-84 is over £2 billion.[1] These figures do not take into account the hidden costs to consumers of many of these policies, nor the difficult-to-quantify economic costs – the opportunity costs in terms of income and growth foregone resulting from the wasteful attraction of additional resources into agriculture by the support system. The external effects, particularly on the environment, must also be taken into account.

---

[1] HM Treasury, *op. cit.*, p. 20. Although part of this expenditure is 'refunded' from Brussels to MAFF, it is paid for mainly out of the UK's budgetary contribution to the EC.

IV. The Economic Arguments for Support

The explanations of the beginnings of agricultural support are historical: depressions, wars and the post-war stimulus to food production at a time of international shortage. The justifications given for its continuation and extension in times of peace, prosperity and abundance of food are different and numerous, although few of the arguments outlined below stand up to close examination.

The 'Special Case'

Many of the economic arguments for support revolve around the idea that agriculture is different from other activities, is in some sense more difficult for producers than other industries, and should therefore be treated as a 'special case'. In the 1960s and 1970s, many groups argued that they were special cases because they were infant industries (electronics), were facing unfair competition (textiles, motor-cycles), or were undergoing painful adaptation (shipbuilding, steel, the coal industry, British Leyland) and would therefore require temporary government aid. The difference between these industries and agriculture is that the protagonists of agricultural support do not regard it as a *temporary* expedient but rather as a *perpetual* requirement.

Special Features of Agricultural Production

The whole subject of agricultural economics has been built on the concept of the 'special case'. Agriculture is different, it is said, because it consists of very large numbers of very small enterprises, none of which is sufficiently large to affect the price of the product – that is, farmers are price-takers – and because it has tended to have a completely separate marketing and processing system dominated by a small number of very large oligopolistic producers (price-makers) who can use their unequal bargaining power to squeeze the farmer. Other alleged problems are the inflexibility of output in relation to price changes; the prevalence of 'trapped resources' (milking parlours and cow cubicles, for example) which are difficult to transfer to other uses and have very low salvage values, and which therefore force the farmer to continue to produce a particular product regardless of market requirements; the very slow capital turnover (three to five years) compared with other industries; and the occurrence of 'lumpy' resources, an example being the necessity to grow, say, 150 acres of cereals in order to justify the purchase of a combine harvester when only a smaller acreage may be available.

Yet the small entrepreneur in manufacturing, retailing and many other service industries is familiar with similar problems and would hotly dispute that any of them are exclusive to agriculture. At this stage in the argument the agricultural economist will produce his trump card: the biological nature of agriculture and the uncertainty it causes the farmer.

Uncertainty

Uncertainty is faced by all entrepreneurs and profit is their reward for handling it successfully, as loss is the penalty for failure. Uncertainty should be distinguished from risk. There is a risk when the statistician can work out the mathematical probability of the occurrence of an event. Risk is something which is insurable (loss from fire, for example), which can be incorporated into the costs of a business, and which therefore does not adversely affect decision-making. Uncertainty is incalculable, characterises situations where there may be a broad range of outcomes, and does affect decision-making. Agriculture, it is argued, faces greater uncertainties than other businesses and farmers' reactions to it can lead to inefficient production. If the uncertainties can be reduced, efficiency will be increased.[1]

Farmers face at least six different types of uncertainty, some of which are common to non-agricultural producers and some peculiar to, or especially prevalent in, agriculture. First, technological uncertainty applies to nearly all economic activities. It is the uncertainty about new techniques, new processes, new types of product, or new machines which may render obsolete existing or planned investment in capital and labour. Secondly, institutional uncertainty is also a common problem which derives from the effects on entrepreneurs of government actions such as changes in investment allowances, interest rates, regional aids, income and consumption taxes, and foreign exchange

---

[1] For a full discussion of uncertainty and its consequences, E. O. Heady, *Economics of Agricultural Production and Resource Use*, Prentice-Hall, USA, 1964, Part III.

policy. Clearly, the more the dependence of an industry on government, the higher is the institutional uncertainty. Thus wholesale government intervention in agricultural prices, output, or incomes can increase this type of uncertainty. The third and fourth types of uncertainty, which relate to the prices of inputs and their availability, apply equally to farming and other activities – except for those farming enterprises, especially livestock, which are highly dependent on other farming enterprises, particularly cereals, for one of their major inputs.

Instability

The two remaining types of uncertainty are those which are commonly held to cause the distinctive instability of agriculture. They stem from the related fluctuations in yields and prices. All farm products are the result of biological processes which are notoriously difficult to control. Land, animals and crops can vary widely in character and quality; they are vulnerable to pests, parasites and diseases; and, above all, they are subject to the climate. It is common for the annual yield of the same crop in the same place to vary by 25 per cent – and variations of 50 per cent or more are not unknown. Plant breeders and agronomists have, however, made immense progress since the war in reducing the yield variability of crops, particularly temperate ones. And most animal production now takes place in controlled environments where – with all credit to the agricultural scientists – considerable precision in controlling yields (for milk, pigmeat, eggs and poultrymeat) has been achieved. The major remaining variability is with tropical primary products where the application of science and management is less and the climate is subject to extreme changes (as the present severe drought affecting much of Africa demonstrates).

Even though variability in yields has been much reduced, there remains the likelihood of some small annual variation in the output of all agricultural products. Because of the very low (and falling) price elasticity of demand for most agricultural products, large price fluctuations will result from very small changes in output. The price elasticity of demand for liquid milk in Britain, for example, is around -0.2 (that is, a 1 per cent reduction in price would lead to an increase in consumption of only 0.2 per cent); the price flexibility, which is measured by its reciprocal, is therefore around 5 (that is, a 1 per cent increase in output would lead to a 5 per cent reduction in price). These figures are for *retail* prices; for prices at the farm-gate, the elasticity would be even lower and the flexibility even higher.[1]

Price fluctuations are also caused and aggravated by the short-run price inelasticity of supply which results from the difficulty or impossibility of adjusting production quickly in response to changes in price. All agricultural production processes take a considerable time – from a few months with some crops, to around two years with beef, and to several years with coffee and rubber. Adjustments to upward changes in prices are therefore inevitably delayed ('lagged'). Adjustments to downward changes also tend to be slow because the variable costs of the farmer tend to be unusually small in relation to his fixed costs. Once specialist buildings, equipment, and machinery have been purchased, it pays to go on producing from them during their life even if product prices are falling. Indeed, the short-run response to price reductions may be 'perverse' – so that output is increased.

There are many different kinds of price and supply fluctuations in agriculture, some of which (like the 'cobweb cycle', of which the 'hog cycle' is an example) are classical cases in economic textbooks.[2] Some of these are now largely historical or theoretical cases because the assumptions on which they are based (for example, that producers do not learn from past experience) no longer hold good. Other types of fluctuation serve useful purposes or are reflections of external circumstances. Examples are long-term fluctuations stretching over many years caused by basic changes in supply and demand; seasonal vari-

---

[1] Elasticity is a technical economic term which measures the response, both in direction and degree, of one economic variable to a (small) change in another, assuming all other factors remain constant. Economists use many different elasticities in their various specialist areas, but the following four are the most widely used in agricultural economics:

  (i) *Price elasticity of demand* measures the percentage change in the quantity of a good demanded in response to a given percentage change in its price;
  (ii) *Income elasticity of demand* measures the percentage change in the quantity of a good demanded in response to a given percentage change in people's incomes;
  (iii) *Cross-elasticity of demand* measures the percentage change in the quantity demanded of good A in response to a given percentage change in the price of good B;
  (iv) *Price elasticity of supply* measures the percentage change in the quantity of a good supplied in response to a given percentage change in its price.

For example, demand is said to be price *inelastic* when there is a less than proportionate change in demand in response to a given change in price (i.e. elasticity is *less* than one); it is said to be price *elastic* when the response is more than proportionate (i.e. elasticity is *more* than one).

[2] For example, G. R. Allen, *Agricultural Marketing Policies*, Blackwell, Oxford, 1959, Ch. 3.

ations in price reconciling demand with seasonally varying production, as with lamb and strawberries; and fluctuations that arise from the general trade cycle, which has, in any case, been less severe and shorter in the post-war than in the pre-war period.

Nevertheless, the supply and demand factors which lead to useless price and income fluctuations are always present. Some commentators, such as Sir Sydney Caine,[1] argue that fluctuations in incomes are not necessarily a bad thing and that it does not matter if a given average income is received at a steady rate or at varying rates over time. It is possible to save during the fat years for the lean years. Farmers, however, have in the past regarded instability as a bad thing and have reacted to it by putting loss avoidance before profit maximisation. They have diversified production (to avoid putting all their eggs in one basket) when specialisation would have been most profitable, have used flexible factors of production where specialist factors would be more efficient, and have avoided investment in capital in order to remain liquid. Survival rather than the optimum use of resources became their goal.

The advocates of guaranteed prices use these arguments to justify their introduction and permanent continuation. But accepting the argument that instability can lead to inefficiency does not inexorably imply that government guaranteed prices are the answer. Reduction of purposeless fluctuations is a desirable goal, but there are many other ways of achieving it which do not involve state interference and all the disadvantages it brings in its train. The provision of market intelligence and the education of producers will reduce 'cobweb'-type cycles. Forward contracting under which the producer agrees to supply specified quantities at specified prices eliminates price uncertainty, as can the use of futures markets which have long existed for grain and more recently for potatoes and pigmeat – and soon for beef in Britain. The formation of co-operatives and groups for buying and selling can increase farmers' bargaining strength and enable them to enter markets in sufficient quantity and with controlled quality to ease the making of long-term contracts. Although it is around 50 years since anything like free markets operated in agriculture in Britain, and we therefore lack recent experience, it is highly unlikely that instability of the kind experienced in the 1920s and 1930s would return in the absence of

---

[1] Sir S. Caine, *Prices for Primary Producers*, Hobart Paper 24, IEA, London, 1966, p. 15.

guaranteed prices, given the present structure of production and marketing.

A final point to be made on instability is that it is not fluctuations in prices but the resulting income instability which is the cause of concern. Stabilising prices when yields vary does not stabilise income: it merely reverses the direction of the income fluctuations arising from price fluctuations.[1] In the old days, when there were few imports to make up shortfalls in home production, farmers used to drink to the toast 'Here's to a bad harvest and a bloody war' because the inelastic demand meant that a drop in production led to a much bigger proportionate increase in price and thus to an increase in revenue. Good harvests led to a drop in revenue. If prices are guaranteed at a constant level, and farmers sell all their production, their revenue will vary in proportion to fluctuations in production. To achieve a stable revenue, prices would have to be varied exactly in proportion to variations in production: a 10 per cent rise in production would be met by a 10 per cent fall in price. A government which proposed to implement such a policy would undoubtedly face the implacable hostility of its farmers.

Low Income

Along with instability, low incomes are the most common justification for government support, guaranteed prices again being the most common remedy. Some agricultural economists have spent a good deal of time investigating and discussing low income,[2] but many, particularly in the basic textbooks, simply take its existence uncritically for granted.[3] Most of the (plentiful) evidence of low income in agriculture is derived from national income and employment statistics and is normally presented as a relative income percentage showing GDP per worker in agriculture as a percentage of GDP per worker in the economy as a whole (or, sometimes, the slightly more sophisticated version showing GDP per worker in agriculture as a percentage

---

[1] For a more detailed discussion, R. G. Lipsey, *An Introduction to Positive Economics*, Weidenfeld & Nicolson, London, 2nd edn., 1966, pp. 139-51.
[2] Especially J. R. Bellerby, *Agriculture and Industry: Relative Income*, Macmillan, London, 1956. Also OECD, *Low Incomes in Agriculture: Problems and Policies*, Paris, 1964.
[3] For example, E. H. Whetham, *The Economic Background to Agricultural Policy*, CUP, Cambridge, 1960.

of GDP per worker in the non-farm sector is used). Thus, if 5 per cent of the GDP originates in agriculture and 10 per cent of the active population is engaged in agriculture, the agricultural relative income is 50 per cent.

Studies by the OECD and other organisations have shown a remarkably constant tendency over long periods for relative agricultural income to be around half that of non-agricultural income. Tables IA and IB reproduce OECD figures for four countries calculated to show relative income percentages for selected years from 1900 to 1962. Table II shows the situation in 17 OECD countries in 1955-56 and 1963-64. Both Tables go a long way towards confirming the 50 per cent norm, with the notable exceptions of Denmark, the Netherlands, and the UK which have high relative income (as does New Zealand). The Tables also show that relative income tends to be very similar both before and after the introduction of support, *and* after support has been considerably increased. The American figure (under measure 2 in Table IB) of 46 per cent in 1900, before the advent of policy, stands at 48 per cent both in 1940 after a decade of policy and in 1962 after over three decades. The Swedish figures are also remarkably constant. The UK figure has tended to remain near parity but did drop to around 50 per cent in the inter-war slump. The more the French have supported their agriculture, the worse relative income has become! Table II tends to show little general improvement in relative income during a period when support was generally increasing. The OECD figures include both farmers' and farm workers' incomes, which are lumped together in their definition. In his classic study, *Agriculture and Industry: Relative Income*,[1] J. R. Bellerby studied the farmer alone and found that, before the war, the worldwide average ratio of 'incentive income' (roughly equivalent to entrepreneurial income) in agriculture to that outside was 55 per cent.

At this stage we can say that there is evidence of long-standing low relative income and that it tends to persist under support policies. This means either that support (of the types already tried) is ineffective in its objective (as stated in the agricultural legislation of many countries) of achieving income parity, or that relative income would have declined even further without support. The reasons for the former view, that support has been unsuccessful and, in the long run,

---

[1] J. R. Bellerby, *op. cit.*

TABLE IA

AGRICULTURAL EMPLOYMENT AND SHARE OF GDP ORIGINATING IN AGRICULTURE IN SELECTED COUNTRIES, 1900 TO 1962

Percentages

Period	USA		Sweden		UK		France	
	a	b	a	b	a	b	a	b
1900	38	22	43	19	8	7	33	n.a.
1910	31	19	41	19	8	6	30	32
1920	27	13	35	15	7	6	28	n.a.
1930	21	10	30	11	6	3	24	21
1940	17	9	27	10	n.a.	4	n.a.	22
1950	12	7	n.a.	7	5	6	20	15
1962	8	4	13	5	4	4	21	9

a=Agricultural employment as percentage of total employment.
b=Percentage of GDP originating in agriculture (at current prices).
Source: OECD, *Agriculture and Economic Growth*, Paris, 1965, p. 101.

TABLE IB

RELATIVE INCOME PERCENTAGES IN SELECTED COUNTRIES, 1900 TO 1962

Percentages

Period	USA		Sweden		UK		France	
	1	2	1	2	1	2	1	2
1900	58	46	44	31	87	86	n.a.	n.a.
1910	61	52	46	33	75	72	106	109
1920	48	40	43	33	85	84	n.a.	n.a.
1930	47	41	37	29	50	48	87	84
1940	53	48	37	30	n.a.	n.a.	n.a.	n.a.
1950	58	55	n.a.	n.a.	120	121	75	71
1962	50	48	38	35	100	100	43	37

1. GDP per worker in agriculture as a percentage of GDP per worker in the economy as a whole:

$$1. = \frac{b}{a} \times 100$$

2. GDP per worker in agriculture as a percentage of GDP per worker in the non-farm sector:

$$2. = \frac{b}{a} \times \frac{100-a}{100-b} \times 100$$

Source: Calculated from Table IA.

TABLE II

ROLE OF AGRICULTURE IN NATIONAL ECONOMIES AND RELATIVE INCOME SITUATION IN 17 COUNTRIES, 1955-56 AND 1963-64

In Percentages

	Average 1955-56		Rel. Inc. 1955-56		Average 1963-64		Rel. Inc. 1963-64	
	a	b	1	2	a	b	1	2
Austria	28·5	14·8	51·9	43·55	20·8	10·6	50·96	45·09
Belgium	9·0	7·6	84·4	80·01	6·3	6·8	107·93	108·46
Canada	17·1	9·4	54·97	50·29	11·5	7·0	60·86	57·87
Denmark	24·9	20·6	82·73	78·17	18·0	14·1	78·3	74·69
France	26·4	10·8	40·90	21·63	19·4	8·2	42·26	37·06
Germany	18·0	7·8	43·33	38·52	11·9	5·0	42·01	38·94
Ireland	38·7	22·1	57·1	44·88	34·0	16·8	49·41	39·18
Italy	39.8	22·2	55·77	14·61	26·4	14·8	56·06	44·45
Japan	39·3	21·3	54·19	41·78	27·4	13·0	47·44	40·94
Netherlands	12·0	11·0	91·66	91·47	8·9	9·1	102·24	85·40
Norway	20·3	6·8	33·49	28·63	16·4	4·8	29·26	29·31
Portugal	46·6	29·7	63·73	54·48	42·4	21·7	51·17	50·30
Spain	44·7	24·7	55·25	49·78	37·2	22·4	60·21	56·47
Turkey	77·4	42·2	54·52	21·31	77·8	39·8	51·15	18·87
EEC (1965)	—	—	—	—	15·9	7·2	45·28	41·02
UK	4·8	4·7	97·91	97·80	3·9	3·6	92·30	91·93
USA	10·9	5·0	45·87	42·98	7·9	3·9	49·36	47·28

(a) Share of active agricultural population in total active population.
(b) Share of total GDP accruing to agriculture (at factor cost and in current prices).

Relative Income:
1. GDP per worker in agriculture as a percentage of GDP per worker in the economy as a whole:

$$1. = \frac{b}{a} \times 100.$$

2. GDP per worker in agriculture as a percentage of GDP per worker in the non-farm sector:

$$2. = \frac{b}{a} \times \frac{100-a}{100-b} \times 100.$$

Source: OECD, *Agricultural Policies in 1966*, Paris, 1967, p. 108.

can never be successful, are discussed in later sections. Next, however, some reservations about this whole approach must be expressed.

Does It Matter?

Many politicians and agricultural economists seem to have assumed that, if farm incomes can be shown to be below average (that is, the relative income percentage is below 100), they are 'low'; and that, if they are low, it is a matter for concern, and therefore for the government to do something about. The links in this chain of argument are weak. First, in a free society (or in any existing Communist one, for that matter) there will always be a range of both pre- and post-tax income brackets. The range will probably be such that the highest income brackets, certainly pre-tax, will be six, eight, or 10 times larger than the lowest ones. What justification is there for saying that agricultural incomes should equal the average? It is accepted that other people's incomes will be below, equal to, or above the average. Why cannot it be accepted for farmers?

Secondly, what is really meant by 'low' income? If low refers to some criterion of need – to some poverty line – the same criterion should apply to people in agriculture as to the rest of the population. Since all the developed countries have welfare states which have comprehensive provisions to relieve poverty, there is no requirement for special provisions for agriculture to alleviate low income in this sense. Only if agricultural incomes were depressed as a consequence of exploitation or oppression against which farmers were defenceless would there be an argument for special government intervention. Leaving aside developing and Communist countries, this is manifestly not the case.

Statistical and Presentational Difficulties

Turning from such fundamental philosophical issues, some qualifications should be made about the uncritical acceptance of national income statistics.

First, it is assumed implicitly in generalised statements about relative incomes that farming is homogeneous, or at least that a similar situation prevails throughout the industry. Yet farming is distinctive above all because of its wide variety. It is impossible to make accurate generalisations about it. There are variations due to geographical area (topography, type of soil, climate); variations between different types of farming (dairying, arable, cattle and sheep, pigs and poultry); variations between farms of different sizes (whether measured by area or size of business); and variations in performance (technical and business management ability) between farmers in any category. The national averages therefore conceal a huge spread of incomes, as a glance at the figures produced annually from the British Farm Management Survey will show. The summary of the survey contained in the 1984 Annual Review White Paper[1] indicates that in 1982-83, for example, net farm income in England ranged from £1,153 for small lowland cattle and sheep farms to £35,424 for large-cropping farms. Within the large-farm category itself, the highest income-type (cropping) still had almost double the income of the lowest (lowland cattle and sheep). And within most groups the 'high-performance' farmers earn 50-100 per cent more than the 'low-performance' ones. Even the most prosperous groups contain significant proportions of loss-makers. Farming contains both very poor and very rich people even after half a century of support. It is clear that the blanket forms of support (which we have had for so long) of prices and inputs for all farmers regardless of income are, from the standpoint of most farmers, let alone that of the rest of the population, totally inappropriate. This contention will be developed in Section V.

Secondly, the British statistics on farmers' incomes have since the 1930s used a concept of income, Net Farm Income (NFI), which itself poses a number of difficulties. NFI is a residue, after all other costs have been paid (including a notional rent for the land), accruing to the farmer and his wife as (a) a reward for their physical labour, (b) a remuneration for their management, and (c) a return on their tenant's capital (that is, their live- and dead-stock and all their capital equipment apart from the land, buildings and fixed equipment). The labour and management rewards could be used for comparisons with people in other jobs. The element of return on tenant's capital, however, has to be excluded if we wish to compare farmers' incomes with those of others employed either inside agriculture (farm workers, farm managers, advisers) or outside (company directors, lecturers, salesmen, doctors, and so on). Even if we compare farmers with other self-employed businessmen whose incomes may also include an

---

[1] *Annual Review of Agriculture 1984*, Cmnd. 9137, HMSO, London, 1984, p. 45.

element of return on capital, that element will be far larger for farming, with its very high capital requirement in relation to turnover, than for most other comparable businesses. Thus net farm income will tend to over-state the income of farmers unless it is adjusted by imputing a reasonable rate of return on tenant's capital to leave a reward for labour and management only. Calculations made at the University College of North Wales[1] in the early 1970s showed that, on most farms, an imputed 10 per cent rate of interest on tenant's capital would normally absorb about one-half of NFI, although the proportion could vary widely from year to year.

NFI can also exaggerate the picture when it contains a large element of valuation change (an increase in the value of animals, stocks of feedingstuffs, fuel, fertilisers, and so on) as it particularly does in times of high inflation. Such changes do not reflect any real increase in the spendable income of the farmer. In 1976-77, stock appreciation accounted for over one-third of NFI. This difficulty led the Ministry of Agriculture first to publish two series of NFI figures, one including and the other excluding stock appreciation, and then to exclude it completely from the White Paper figures for what is now termed 'Farming Income'.

Total Income Understated

On the other hand, there are several aspects of NFI which tend to *under*state the total income of the farmer, and even more so his standard of living or economic well-being. In the first place, NFI is a measure of the income the farmer derives from farming as such and does not measure his income, or that of his wife, from other sources which may be on or off the farm. Although everyone connected with agriculture has long known that farmers frequently have other employment or other businesses, detailed information about them (for Britain, at any rate) has been scanty. Recent published research by Dr Berkeley Hill of Wye College[2] has shed further light on this important but neglected aspect.

---

[1] K. A. Preston, *Farm Incomes: an Examination of Farm Management Survey Data* (unpublished dissertation), Dept. of Agric., UCNW, Bangor, June 1975.

[2] B. Hill, 'Concepts and Measurement of the Incomes, Wealth and Economic Well-Being of Farmers', *J. Agric. Econ.*, Vol. xxxiii, No. 3, September 1982, and 'Farm Incomes: Myths and Perspectives', *Lloyds Bank Review*, July 1983.

The other occupations engaged in by farmers include, for example, fishing, retailing, road haulage, factory work, joinery, and undertaking. Many have businesses more closely connected with farming, such as agricultural contracting and merchanting, fencing, and food shops. Some farmers have other jobs which are highly 'up-market'; they may be Lloyds underwriters, stockbrokers, barristers, and even Cabinet Ministers. On the farm itself it is often possible to earn additional income from bed-and-breakfast guests, caravan and tent pitches, farm buildings converted into rentable accommodation, pony trekking, and fishing and shooting rights.

Although it is not known how much income farmers gain from each of these many sources, it must in total be considerable. As Hill points out, almost a quarter of all UK farmers (23 per cent) have an additional source of earned income; in other EC countries the proportion is even higher (Germany 43 per cent, Italy 30 per cent, and for the Community as a whole, excluding Greece, 27 per cent). To 'other earned income' must be added wives' income, investment income, and other income. Figures from the Inland Revenue Survey of Personal Incomes, quoted by Hill and reproduced in Table III, illustrate how diverse are the sources of income of those classified as having agriculture as their major source of self-employment income. In 1978-79, farmers and their wives derived only 63 per cent of their total income mainly from their farms, the remainder coming from other employment and other incomes (20 per cent) or from investments (17 per cent).

It is plainly nonsensical to use farm income alone as evidence that farmers are absolutely or relatively poor, and to base policy on it. If other sources of income are taken into account, the picture may look very different. Even the most diehard protagonist of support would find it hard to accept the case for subsidising, in times of a glut of food, the incomes of farmers for whom farming is a secondary occupation (be they a German car-worker or a British parliamentarian), or who are earning large profits from the use of their farms for tourist purposes.[1]

---

[1] An unpublished dissertation (D. H. Chavasse, *Farm Tourism in England and Wales*, Dept. of Agric., UCNW, Bangor, May 1984) shows real internal rates of return as high as 69·8 per cent for static tourist-owned caravan sites; 64·6 per cent for large touring caravan and tent sites; and 32 per cent for farmhouse bed-and-breakfast catering.

TABLE III

COMPOSITION OF TOTAL INCOME OF COUPLES AND
INDIVIDUALS IN THE TRADE GROUP AGRICULTURE OF
THE 1978-79 SURVEY OF PERSONAL INCOMES

	Percentage of total income %
Self-employment income (i.e., mainly from the farm)	
Husbands (and single persons)	54
Wives	9
Employment income (i.e., as hired employees)	
Husbands	8
Wives	7
Other (including pensions)	5
Earned income	83
Rents	2
Building Society interest	5
Other interest and dividends	11
Investment income	17
ALL INCOME	100

Note: 'Investment income' total does not add up correctly due to rounding.
Source: Berkeley Hill, 'Farm Incomes; Myths and Perspectives', *Lloyds Bank Review*, London, July 1983, p. 44.

Standard of Living

Even if we concentrate on purely farming income, a distinction should be made between income (as represented by, say, NFI) and standard of living, because it is quite possible for two people to have an identical declared (taxable) income and yet enjoy widely different standards of living, regardless of other income sources. This situation can arise legally or illegally. Farmers can make unrecorded sales of produce, which are difficult for the authorities to detect (particularly when the sales take place between farmers), and therefore enjoy a much higher standard of living than their apparent income would suggest. There is no doubt that such practices occur. Without further evidence, however, it would be unfair to single out farmers from other businessmen in this respect.

Quite legitimately, nevertheless, the farmer does enjoy advantages from his farm which are not available to the rest of the population. The farm can provide benefits in kind such as milk, meat, eggs, potatoes and other produce which can be charged at less than half of the shop price – though the variety of produce from most farms is much less today that it used to be. There is usually also a plentiful supply of timber for fuel available free on the farm. The farmhouse normally comes with the farm and is paid for in the same cash purchase, mortgage, or rent payment as the farm. It will frequently be of larger size, higher quality, and better character than the houses owned by other people on similar incomes. Since the farmhouse is normally on the same site as the farm buildings and adjoining land, there is no cost (nor time lost) in travelling to work, which is a significant item of expenditure out of taxed income for most people. Many farms provide opportunities for fishing and shooting which would otherwise have to be paid for dearly.

In addition, the farmer enjoys the advantages the taxation system allows self-employed people compared with employees. He can run mainly out of the business any type of car or vehicle he chooses, and is entitled to generous allowances for a telephone, electricity, and other fuel. It is difficult to put a figure on all these perquisites, many of which affect the quality of life as well as the material standard of living, but it is reasonable to assume that the value to a medium-sized farmer would be of the order of £4,000-6,000 a year in terms of the pre-tax income which an employed person living in an urban area would have to earn to enjoy a similar standard of living. In other words, a £10,000-a-year farmer might well be enjoying a standard of living equivalent to that of a £16,000-a-year sales manager or civil servant. And, at the bottom end of the income scale, it is much easier for a farm family than for a poor urban family to survive on its own produce and to be housed – even with virtually no cash income.

Economic Well-being

We must go further than either money income or standard of living, however, to obtain a full picture of a person's economic well-being.

This task requires an examination of his assets and unrealised capital gains. In both these respects the owner-occupier farmer and the landowner are in a much stronger financial position than their income would imply. The same cannot, however, be said of the tenant farmer who does not own land, which has appreciated in value more than any other component of a farmer's capital such that an owner-occupier of 400 acres in Britain is now a millionaire. Hill's article in the *Journal of Agricultural Economics* (1982)[1] discusses both these aspects and provides estimates based on 1976-77 Farm Management Survey data, which show an average real gain in land values over the year of £347 per hectare. Over the period 1970 to 1977, the average real annual capital gain was £120 per hectare (at 1977 prices), which would have added almost 50 per cent to average NFI in 1976-77 and raised the proportion of farms with annual incomes of £10,000 and over from 45 to 65 per cent. Of course, capital gains pose problems of measurement, can fluctuate widely, and are frequently difficult to realise. It would be inconvenient for a farmer to keep selling off parcels of his land to maintain it at a constant total value. But as Hill observes:

'Unrealised gains represent the value of rights which the owner *might have* exercised in consumption without diminishing the value of his wealth'.

The high asset values in farming must obviously be taken into account in assessing the economic status of the farmer. Most people would not agree that millionaires should be subsidised even if their cash income was very low. The assets of the farmer could be converted into a high income-flow if used in a different manner – by purchasing an annuity, for example. Hill gives figures derived from converting estimates of farmers' net worth into an annual income-flow using an annuitisation formula. Such exercises have to be accepted with caution since they depend heavily upon the choice of interest rates and asset values. But using a conservative 2 per cent real rate of return and valuing the land at the low tenanted values, he estimates that the income would be one-third larger than the existing all-farm NFI.

It can thus be seen that bringing capital gains and annuitised asset values into the picture does raise very considerably the (potential) income of the farmer. These aspects are especially relevant when the farm lobby is crying out for an increase in the already heavy support

---

[1] B. Hill, *op. cit.*

it receives. They are, however, less relevant to the arguments for moving from a support to a free-market policy. The large capital gains and high asset values are mainly the consequences of support and did not exist in the past when (as in the 1920s and 1930s) there was little or no support. Their relevance today has more to do with the capital losses farmers would suffer if support were reduced or abolished and the extent to which farmers could or ought to be compensated for such losses as a result of a change in direction by government.

The Demand for Food

Moving on from the supply side of agriculture and the income of the farmer, we come to the demand side. Here are to be found some of the major sources of the downward pressure on farm income present both before and after the advent of support, and notwithstanding the factors which can lead to an understatement of the true income. The characteristics of the demand for food have long been understood by economists even though many farmers and their sympathisers wilfully choose to ignore them because of their unpalatable implications. The two relevant aspects are the relationship between the demand for food and (a) income (income elasticity of demand) and (b) its price (price elasticity of demand). Since these are discussed at length in many agricultural economics textbooks,[1] only the essentials require to be examined here.

As in so many other matters, Adam Smith in 1776 had the last word – oft quoted and incomparably expressed – on the income relationship:

'The rich man consumes no more food than his poor neighbour. In quality it may be very different, and to select and prepare it may require more labour and art; but in quantity it is very nearly the same. But compare the spacious palace and great wardrobe of the one, with the hovel and few rags of the other, and you will be sensible that the difference between their cloathing, lodging, and houshold furniture, is almost as great in quantity as it is in quality. The desire of food is limited in everyman by the narrow capacity of the human stomach; but the desire of the conveniencies and

---

[1] For example, J. W. Barker, *Agricultural Marketing*, OUP, Oxford, 1981; B. E. Hill and K. A. Ingersent, *An Economic Analysis of Agriculture*, Heinemann, London, 1977; G. Hallett, *The Economics of Agricultural Policy*, Blackwell, Oxford, 2nd edn., 1981.

ornaments of building, dress, equipage, and houshold furniture, seems to have no limit or certain boundary.'[1]

We may have progressed from the era of hovels and rags, but the anatomical truth about the 'narrow capacity of the human stomach' will always be valid. The demand for food remains relatively constant from month to month and from year to year. It is a stable and universal demand which is highly unlikely to disappear or to change significantly. All these factors are to the advantage of the farmer, unless at some time far in the future synthetic substitutes replace food from plants and animals. But it is a demand which, at the high incomes we have attained, grows extremely slowly in relation to income. This matter was investigated by the 19th-century German statistician, Ernst Engel, who surveyed households in Saxony and Belgium according to their income brackets. His conclusions have come to be known as 'Engel's Law', one of the firmest of economic laws, which states that expenditure on food as a proportion of total expenditure declines as income increases, or, to put it more technically, the income elasticity of demand for food declines as income increases. The law may be verified by cross-section analysis of households at a given time, or by time-series analysis relating changes in expenditure on food to changes in income *over* time.

Trend of consumption elasticities

Economists think not only of income elasticities of expenditure on food (which may be measured at the retail stage or at the 'farm-gate') but also of income elasticities of consumption where changes in physical consumption (expressed by weight or volume, or perhaps in terms of calories, fats, or animal protein) are related to changes in income. Consumption elasticities tend to be lower than expenditure elasticities, and for many food products – and for food in aggregate – can be zero or even negative at high incomes. The pattern of consumption of the individual, who will tend to satiate himself when first moving from a state of hunger and then gradually reduce his consumption to a somewhat lower and more comfortable level, tends to be repeated for countries as a whole. The over-eating and obesity of the

---

[1] Adam Smith, *The Wealth of Nations*, ed. Edwin Cannan, Methuen, London, 1950, Vol. I, Bk. I, Ch. XI, p. 183.

TABLE IV

ANNUAL FOOD CONSUMPTION PER HEAD IN THE UNITED KINGDOM, 1955 TO 1977

	Quantities (kg) per head per year				
	1955	1965	1970	1976	1977
Liquid milk (litres)	143·6	141·7	137·2	140·4	135·4
Cheese	4·1	4·6	5·4	6·1	5·5
Beef and veal	21·5	20·8	22·2	21·1	21·1
Mutton and lamb	11·1	10·5	9·6	7·6	7·0
Pork	8·1	11·7	11·1	10·2	11·5
Bacon and ham	11·2	11·7	11·4	8·5	8·9
Poultry	2·9	7·6	10·7	11·5	11·4
Fish	9·7	9·4	8·9	8·4	7·4
Eggs (number)	229	271	275	248	248
Butter	6·7	8·8	8·8	8·3	7·8
Margarine	8·0	4·4	5·0	5·4	5·4
Sugar	46·2	45·4	44·0	38·9	38·6
Fruit	51·3	55·4	56·6	56·2	52·9
Potatoes	106·2	100·9	103·5	85·0	96·6
Other vegetables	55·2	60·9	63·0	65·9	65·6
Wheat flour	82·8	70·3	66·2	66·2	65·3
Tea	4·2	4·1	3·9	3·6	3·2
Coffee	0·6	1·3	2·0	2·1	1·7
Beer (litres)	79·7	91·4	101·5	118·9	117·8
Wine (litres)	1·5	2·9	3·8	6·7	6·4
Spirits (proof litres)	1·0	1·4	1·6	2·9	2·5

Source: *Farming and the Nation*, Cmnd. 7458, HMSO, London, February 1979, p. 9.

1950s and 1960s have been replaced by the diet-and-fitness consciousness of the 1970s and 1980s, first in America and now in Europe.

Such a trend of consumption in the UK is illustrated in Table IV. Out of the 16 food items in the Table, consumption of all but five fell between the mid-1950s and the late-1970s – and for four of those five was fairly static. The only major increases were in coffee, poultry (which has replaced other meats), and alcoholic beverages (from providing the grain and grape content of which farmers can gain some consolation). In contrast to most other developed countries, consumption of potatoes has held up remarkably well – no doubt as a result of the average Briton's desire for 'chips with everything'. But total consumption of calories declined from 3,170 per head per day in 1955 to 2,920 in 1977, around which figure it has stabilised. Consumption per head of fats, vegetable protein and carbohydrates fell over that period, whilst consumption of animal protein rose to a peak in the early 1970s and has since fallen back a little.[1]

Trend of growth in demand

Growth in the demand for food (D) is a function of three variables: the rate of population growth (p), the rate of growth of disposable income (g), and the income elasticity of demand (n). It may be expressed in the simple equation:

$$D = p + ng$$

The rate of growth of population, like income elasticity, tends to decline as income rises. Rich people have fewer children than poor people. Thus population growth is now generally below 1 per cent a year in developed countries (0·6 per cent a year in the UK and 0·8 per cent in Western Europe as a whole was forecast for 1970 to 1980)[2] and makes only a small contribution to the growth of demand for food. (In developing countries, where the opposite applies, population growth is of the order of 2 to 3 per cent a year.) Similarly, ng is very small because, even if economic growth (g) is high, which it has not been over the past decade, the income elasticity (n) is so low as to make ng unlikely to exceed 0·5 per cent – and in recent years it has been much lower. If g is zero or negative, ng must also be zero or negative – in which case the growth of demand for food would be solely dependent on population growth. Even with an optimistic assumption of economic growth at, say, 4 per cent, and assuming population growth at 0·6 per cent and income elasticity of expenditure at the farm-gate of 0·1 per cent, the equation produced is:

$$D = 0·6 + (0·1 \times 4·0) = 1·0 \text{ per cent a year.}$$

---

[1] The statistics are from *Farming and the Nation*, Cmnd. 7458, HMSO, London, February 1979, pp. 9 and 10.

[2] Interesting statistics and worldwide projections of demand for food, from which these figures are taken, are given in W. J. Thomas, 'Looking at the future of agriculture', *J. Agric. Econ.*, Vol. xxiv, No. 3, September 1973. The most recent figures of population growth (*The Economist Diary 1985*, p. 40) are even lower. From 1972 to 1983 the UK annual rate was 0·1 per cent, for the EC (10 countries) 0·3 per cent, for France 0·5 per cent, and for West Germany zero.

Thus, in the foreseeable future, the rate of growth of demand for food at the farm-gate is highly unlikely to exceed 1 per cent a year in Britain unless some major unexpected change takes place. In the EC and the USA, it is highly unlikely to exceed 1·5 to 2·0 per cent a year. These are inescapable facts which have to be faced by farmers and policy-makers alike.

Indeed, from the farmers' point of view the situation is likely to become even worse because the farm-gate income elasticity is continuing to decline even more rapidly than the retail elasticity as the other trend discerned by Adam Smith goes on. That is, as we become richer, we require more 'labour and art' to 'select and prepare' our food. The services attached to our food (grading, pre-preparation, packaging, freezing, and so on) are provided by the food marketing industry which comprises the food manufacturers, wholesalers, and retailers. In Britain this is a highly competitive and efficient industry which has become bigger than agriculture itself in terms of the value of its output; its rate of growth is also higher than that of agriculture. In advanced countries, the proportion of expenditure on food which goes into processing and distribution (known as the 'marketing margin' and varying considerably from product to product) may be as high as 60 per cent in aggregate (as in the USA). Thus only 40 per cent returns to farmers (the 'producers' share'). In the UK, the marketing margin (at current prices) was 52·5 per cent in 1971, rising to 56·3 per cent in 1980[1] – which meant that out of every £1 spent on food the farmer received just under 44p.

Expenditure on food as a proportion of consumers' total expenditure (which may be 60 per cent or more in developing countries) had in Britain fallen to 30 per cent by the early-1960s, to just over 20 per cent in the early-1970s, and is estimated to have been 17·1 per cent in 1983.[2] The income elasticity of expenditure at the retail stage has fallen from 0·4 in the late-1950s to below 0·2. The farm-gate elasticity, reflecting the falling producer's share, must now be well below 0·1.

Constraint on growth of farming industry

These factors impose a constraint on the growth of the farming industry. Certain products have higher income elasticities than others –

---

[1] Source: *Annual Abstract of Statistics 1983*, HMSO, London, 1983, p. 207.
[2] Sources: Cmnd. 9137, *op. cit.*, p. 14, and G. Hallett, *op. cit.*, p. 125.

which may range at the retail stage from 1·16 for fresh cream (a luxury product) to 0·03 for liquid milk and eggs (basic products), and to negative figures for sugar, margarine, fresh potatoes and bread.[1] It may thus be possible for some sectors of farming to expand more rapidly than others, but the expansion of one will be at the expense of another. No sector faces anything approaching unit elasticity (where a 1 per cent increase in income leads to a 1 per cent increase in expenditure on the product). And farmers find it hard to accept that expenditure on their product grows at a negligible rate whilst, as Adam Smith observed, there seems to be 'no limit' to the growth of expenditure on housing, clothing, and 'equipage'.

The only way that agriculture in a wealthy country can experience a growth in expenditure on its products in line with the growth of expenditure generally is by replacing imports or by exporting. If it does either or both of these, it may maintain its expansion – but only at the expense of similar products produced by other countries. In this respect Britain's farmers have been fortunate in having the scope to replace a massive amount of imported food, which was still one-half of our total supplies in the late-1950s. They have been able to do this steadily until, by 1983, they were producing an estimated 60 per cent of the value of all food (including tropical products) consumed in the UK – equivalent to 76 per cent of all indigenous-type food (that is, those products which are grown here commercially in significant quantities).[2] Britain has now almost reached or passed self-sufficiency in many products, including cereals, beef, pork, poultrymeat, and eggs. Our self-sufficiency in butter has increased from 18 per cent in the early-1970s to 66 per cent in 1983 (estimated); in milk powder from 81 to 800 per cent, whilst we have always been self-sufficient in liquid milk.[3] In the EC as a whole, self-sufficiency in most temperate products (wheat, barley, sugar, dairy products, eggs, pigmeat, poultry-meat) has long been passed, and the acute problems of surplus (discussed below) are well known.

Even with constant prices for farm products (relative to other prices), therefore, the revenue of the agricultural sector will grow much less rapidly than that of many other sectors and of the economy as a whole.

---

[1] J. W. Barker, *op. cit.* p. 26.
[2] Source: Cmnd. 9137, *op. cit.*, p. 14.
[3] *Ibid.*, pp. 23-36.

This will mean that the relative income (from farming) of an unchanging number of farmers will decline unless they are able to cut their average unit costs of production sufficiently rapidly to maintain the growth of their profits (net income) in line with general economic growth. Unfortunately for the farmers, most of the reductions in costs (increases in productivity) have been achieved over the years by adopting new technologies (mechanical, chemical, genetic) which have raised the output of existing amounts of resources (land, animals, labour) – frequently, though not invariably, by exploiting economies of scale. Although the total agricultural area of Britain has declined slightly over the past century, the number of cattle on it has more than doubled to over 13 million head, the number of pigs has more than trebled, whilst the number of sheep has remained fairly constant. Not only have the numbers of livestock risen, but the average yield per animal has also increased considerably – in milk, the yield per cow has doubled since the Second World War. On the cropping side, the yield increases have been even more spectacular with a trebling of cereal yields since the First World War and a doubling since the Second. Crop production on a similar arable area has rocketed: wheat production is up from 1·6 million tons in 1914 to over 10 million now, and barley from 1·2 million to 8 million tons.[1] These figures explain the enormous increases in self-sufficiency mentioned above and are illustrative of the vast worldwide capacity to expand the output of temperate agricultural products far faster than the growth in demand.

Whenever the increase in supply exceeds the increase in demand, farmers become victims of their own success because the other major feature of the demand for food, its low price elasticity, comes into play. Price elasticity has already been mentioned in connection with price fluctuations, which are accentuated by low elasticity.

Income and substitution effects

Price elasticity of demand is determined by two factors known as the income effect and the substitution effect. The income effect depends upon the proportion of income which is spent on a particular product; the larger the proportion of income it absorbs, the larger will be the

---

[1] Figures derived from Ministry of Agriculture, Fisheries and Food (MAFF), *A Century of Agricultural Statistics: Great Britain 1866–1966*, HMSO, London, 1968, and Cmnd. 9137, *op. cit.*

responsiveness of the consumer to changes in the price of the product, and conversely. As was observed with income elasticity, the proportion of income spent on food as a whole is now small (17 per cent in Britain); and on individual food items such as milk it is very small (1 to 2 per cent), and getting smaller with rising incomes. The substitution effect depends upon the willingness and ability of consumers to switch into or out of substitutes as the price of a product changes. There is obviously no substitute for food as a whole, although the quantity of food consumed can be altered within limits. But the consumer can and does shift markedly from one foodstuff to another similar one in response to relative food prices. There are, therefore, high cross-elasticities within certain categories of foodstuffs which can lead to high price elasticities for individual commodities (such as -1.65 for pork and -1.37 for mutton and lamb). For groups of foods (such as all carcase meat at -0.23) and for individual items which have no close substitutes, however, the elasticities are all below unity, and in many instances very low (eggs -0.09, liquid milk -0.16, and fresh potatoes -0.17).[1]

The combined result of the two effects is that the price elasticity of demand for food, like income elasticity, is now very low and declining. The problem which results from the associated reciprocal, the high price flexibility, is that small output changes lead to large price changes. Whenever price elasticity is below unity and price flexibility thus above unity, any excess of supply growth over demand growth will, in the absence of intervention, lead to a more than proportionate reduction in product prices and thereby a reduction in the revenue of the producer. If the price elasticity is -0.25 ($\frac{1}{4}$), which is quite typical for farm products, the price flexibility will be -4. A 1 per cent excess supply will lead to a 4 per cent fall in the price of the product which will reduce total revenue by 3·04 per cent.

If there is a persistent tendency for the growth of supply to exceed that of demand, as there is in agriculture, there will be a persistent downward pressure on farm product prices and on the industry's total revenue which will call for even bigger reductions in costs per unit of output to maintain farmers' incomes than those required as a consequence of the low income elasticity. Farmers will appear to be in the jaws of a nutcracker exerting pressure on them from two

---

[1] J. W. Barker, op. cit., p. 26.

directions simultaneously. The growth in revenue from their sales will be restricted by both the slow rate of growth in the volume of sales and by the downward trend in agricultural prices relative to non-agricultural prices. Such a trend has been called the *'terms of trade model'*. This has been expounded by agricultural economists such as Gavin McCrone[1] who highlight the theoretical and actual decline in the terms of trade between agriculture and other sectors, which requires farmers to offer an ever larger quantity of agricultural products to purchase a given quantity of industrial goods.

The Farm Problem

Demand factors constitute some of the main ingredients in what has come to be known as the 'farm problem', a term used by American agricultural economists[2] who took the lead in the post-war period in analysing the difficulties of agriculture. To describe this problem fully, it is necessary to return to the supply side of agriculture and to examine further the reasons for the tendency to excess supply. The spectacular improvements in *technical* productivity, already referred to, provide a partial explanation. It does not matter at this stage whether *economic* productivity has improved as dramatically as the farm lobby would have us believe, but there is no doubt that *physical* output per acre of crops, grass, meat, and milk has made spectacular strides, as has the output of meat and milk per animal. Not only is each acre and each animal producing more; the number of acres cultivated and harvested and the number of animals fed, milked, and tended by each man employed (that is, physical product per man) have increased dramatically. Such is the triumph referred to by Richard Body in *The Triumph and the Shame*. It has been achieved by plant and animal scientists, chemists, and geneticists simultaneously with a mechanical revolution,[3] a man-

---

[1] G. McCrone, *The Economics of Subsidising Agriculture*, Allen and Unwin, London, 1962, Ch. 8.

[2] It is difficult to know who first drew attention to the farm problem, but an important early contribution was T. W. Schultz, *Agriculture in an Unstable Economy*, McGraw Hill, New York, 1946. Also worthy of mention are: G. S. Shepherd, *Farm Policy: New Directions*, Iowa State University Press, Ames, 1964; and E. O. Heady et. al., *Roots of the Farm Problem*, Iowa State UP, Ames, 1965.

[3] Wholesale mechanisation is a much more recent phenomenon than is commonly believed: E. H. Whetham, 'The Mechanisation of British Farming 1910-1945', *J. Agric. Econ.*, Vol. xxi, No. 3, September 1970, pp. 317-326.

agement revolution, and now an electronics revolution. Not only have new technologies become available to the farmer but improvements in education, information and advice have meant that the dissemination, acceptance and take-up of new ideas have broadened and accelerated.[1]

The consequence of all these factors has been that the gross output (total quantity of produce sold off the 'national farm' valued at constant prices) per worker employed in agriculture (gross productivity) has risen in the post-war period at compound rates of growth of the order of 4–7 per cent a year in the UK, the USA and Western Europe. Since, at the same time, the agricultural labour force has been declining by 2–4 per cent a year, the volume of output has been increasing by around 2–3 per cent a year. In the decade from 1950 to 1960, for example, gross output in the UK, West Germany and the Netherlands rose by 2·7 per cent a year, in France and Italy by 2·8 per cent, and in the USA by 1·8 per cent.[2] There is thus a tendency for a growth in supply of, say, 2 per cent to exceed the growth in demand by one percentage point a year – which could lead to a depression of agricultural prices, in the absence of government intervention, of the order of 4 per cent a year. Of course, this is not a trend which would necessarily persist at the same rate every year. There are always unforeseen movements in international demand and supply caused by wars, disease, droughts, floods, and famines. As in the early 1970s, there may be quite long interruptions in the trend. But the underlying pressures are always there. These pressures are summarised in Table V which brings together the ingredients of the farm problem.

It is on this 'problem', or more accurately certain components of it, that protagonists of agricultural support base many of their arguments for long-term intervention in agricultural markets. They maintain that, since there is a long-run downward pressure on farm product prices and revenues, there is consequent pressure on farmers' incomes; and that this should be tackled by artificially raising product prices (directly or indirectly by restricting supply), by subsidising the costs of inputs, or by directly subsidising farmers' incomes – or perhaps by all three methods at once. This is certainly how farmers' organisations

---

[1] This development is discussed at more length in Open University, *Decision Making in Britain III, Parts 1–5: Agriculture*, Open University Press, Milton Keynes, 1975, Part 2.

[2] G. Hallett, *op. cit.*, p. 107.

TABLE V

ILLUSTRATION OF THE FARM PROBLEM IN THE POST-WAR PERIOD

(*Typical figures applicable to Western Europe and the USA*)

	Percentage rate per year %
Increase in Demand for Farm Products (dependent upon population growth, economic growth, and income elasticity)	+1
Increase in gross output per person employed in agriculture (gross productivity)	+4
Decrease in persons employed in agriculture (rate of exodus)	−2
Increase in Supply of Farm Products (4% − 2%)	+2
Excess of Supply Growth over Demand Growth (2% − 1%)	+1
Depression of Agricultural Prices (assuming price flexibility of −4)	−4
Depression of Agricultural Revenue	−3·04

everywhere view the position. And governments in every OECD country seem to have concurred in this view by implementing such policies (perhaps because it appeared to be politically necessary), even if they no longer accept the analysis.

Certainly, few agricultural economists today accept that high support prices, supply controls or input subsidies are an efficient, effective, or equitable solution to the 'problem', although many still refer to a 'farm income problem'.[1] The American agricultural economists have shown convincingly[2] that, if there *is* a 'farm problem', it is not a price problem, it is not a supply problem, and it is not an income problem. All three are merely readily discernible outward manifestations, or symptoms, of the true problem.

Table V suggests three remaining variables which could possibly

---

[1] For example, Hill and Ingersent, *op. cit.*, Ch. 4.
[2] This viewpoint is expressed forcefully in G. S. Shepherd, *op. cit.*

lie at the heart of the problem: the demand increase, the gross productivity increase, and the rate of exodus. Demand could possibly be raised by subsidising sales, which has been tried occasionally but is very costly and ineffective given the low price elasticity. The only major increase in demand comes from a growth in population, but few would advocate the encouragement of a population explosion to assist farmers' incomes! Gross productivity growth could be slowed down by stopping research and development. Again, it would be ridiculous to prevent commercial technical improvements except where they may have detrimental side-effects on human health (pesticides, fungicides, and other chemicals may possibly do this) and on the environment. But the case for state-subsidised research in temperate agriculture does look increasingly weak in the context of over-supply, particularly when many further developments are likely to be in the realm of diminishing returns and will probably be economically unviable any way. There is evidence that the high rate of technical progress of the past three decades may already have peaked in several countries and has now fallen off, although the scope for the adoption of existing technology remains vast in the Third World, which has the potential to increase output at least fourfold.

Nevertheless, farmers in the developed countries remain on a 'treadmill' of technical innovation which constantly shifts the agricultural supply curve upwards and to the right. The 'treadmill theory' was developed by Professor Willard Cochrane[1] to show how, in raising their output, farmers unwittingly reduce the prices of their products. New techniques are not adopted by all farmers at once. First, the small number of 'innovators' pioneer a new technique and profit from it (sometimes, of course, they lose if the technique is unsuccessful). Then the 'early adopters' take it up and they also benefit. By now, however, the supply curve is already shifting upwards as the 'late adopters' and finally 'the laggards' complete the adoption process. By this stage prices have fallen and the race is again on to find new techniques and to repeat the whole process. As the rates of technical progress and adoption increase, so the treadmill turns more rapidly. Price support is obviously no solution to the treadmill problem; rather, it will aggravate it by simply adding to the incentive to expand

---

[1] W. W. Cochrane, *Farm Prices: Myth and Reality*, University of Minnesota Press, Minneapolis, 1958.

output. Support costs will rise until, ultimately, they become politically unbearable and support has to be stabilised or reduced.

The Rate of Exodus

We are left with the rate of exodus as the final possible solution or cause, or both. It is this factor, the supply of labour, which is at the heart of the farm problem. In Table V, the growth of supply would match the growth of demand if the rate of exodus was 3 per cent instead of 2 per cent. The question is: 'Why does the rate of exodus not adjust itself so as to reduce the pressure on farm incomes?' This is the 'adjustment problem' to which there is no single answer but a variety depending on the time, the place, and the circumstances.

The movement of labour out of agriculture is one of the key features of economic progress and has been going on for at least two centuries. By the mid-19th century only about one-fifth of the working population of Britain was employed in agriculture; by the turn of the century the proportion was down to 8 per cent, and by 1980 it had fallen to 2·6 per cent. But the bulk of this exodus comprised hired workers as opposed to farmers themselves. The estimated number of farmers in England and Wales was 249,000 in 1851, falling to 229,000 in 1911; it then rose to 248,000 in 1931, 270,000 in 1947, and reached a peak of 306,000 in 1961; by 1978 the figure had fallen back again to 234,000. Meanwhile the number of full-time workers (including family workers), which stood at 1,380,000 in 1851, fell steadily from 803,000 in 1911 to 620,000 in 1931, to 549,000 in 1947, and to only 177,000 in 1978.[1] The ratio of hired workers to farmers thus fell from 5·5 in 1851 to 2·5 in 1931, to 2·0 in 1947, and to only 0·75 in 1978.

Fortunately for British agriculture, in contrast with most of the Continent and America, large numbers of employees were able to take the brunt of the 'adjustment' process. Either they were pushed out of agriculture by their employers during hard times, as in the two depressions; or they responded to the 'pull factors' of alternative employment, as in the post-war period. Econometric studies[2] have shown that the British agricultural labour market has responded in an

---

[1] G. Hallett, *op. cit.*, p. 235.
[2] Referred to in D. Metcalf, *The Economics of Agriculture*, Penguin Books, London, 1969, pp. 44-46.

economically rational manner to such factors as the height and trend of industrial unemployment, the degree of regional industrialisation and the proximity of alternative employment, and the ratio of agricultural to industrial earnings. Thus the rate of exodus from agriculture of full-time hired workers during the full-employment conditions of the 1960s was around 5 per cent a year, and was 4 per cent even in the more economically troubled decade from 1972 to 1982.[1] But the UK now has only 167,000 (1983 provisional figure) regular, whole-time farm-workers, and the scope for further adjustment to their numbers is small.

It is from the farmers that most of any further decline will have to come, and they are a very different kettle of fish from their workers in that they appear not to respond to economic forces anything like so readily. From the British experience, expressed in the figures above, we might expect a long-run decline in the number of farmers of about 1 per cent a year. Experience elsewhere shows a similar propensity to leave the industry. Agricultural economists and rural sociologists have spent much of their time trying to explain what they call the farmer's 'immobility'. Bellerby's pioneering work[2] provoked much subsequent debate resulting in two schools of thought which appear to be contradictory but which each contain part of the answer.

Asset-fixity Theory

The first school promotes the 'asset-fixity' theory, arguing that farmers and much of their capital are 'trapped' resources which are locked into agriculture and cannot escape from it without substantial losses. This idea was advanced by another American agricultural economist, Professor Glenn L. Johnson.[3] The theory depends on the notion that there is a major difference between the acquisition costs and the salvage values of durable capital assets, the extreme cases of which are specialist buildings which cannot be sold without being demolished and transported at high cost. As mentioned above (pp. 19-20), this idea is a partial explanation of why agricultural supply is so slow to contract when

---

[1] Source: Cmnd. 9137, *op. cit.*, p. 5.
[2] J. R. Bellerby, *op. cit.*
[3] G. L. Johnson, 'The Labour Utilisation Problem in European and American Agriculture', *J. Agric. Econ.*, Vol. xiv, No. 1, 1960.

prices are falling; the marginal value product of assets is less than their purchase price but still larger than their sale or salvage value. It is, however, the notion that the farmer himself is in an analogous position[1] which is most relevant to the present argument.

Johnson maintains that the farmer frequently has a very low 'salvage price' – that is, the price he could command for his services outside agriculture plus the value of (or income from) his saleable assets currently invested in farming less the costs of transferring out of agriculture. If the farmer is old or middle-aged he probably lacks experience of other work – even if it were available in his area, which it frequently is not, and even if employers are willing to employ him, which they frequently are not, except in the most menial of jobs. To gain another job he would have to sacrifice his pride and his independence. If he is also a tenant or has heavy borrowings on his land, the net capital on selling up may be quite small, certainly after purchasing another home. The costs of transfer in agents', auctioneers', and lawyers' fees, and in removal expenses, will be considerable. There is no incentive for him to move out so long as his 'market price' – his net cash family income – exceeds his salvage price. Indeed, he may still stay on if his market price is lower than his salvage price. He may value the non-monetary (psychic) benefits of farming and decide that they are sufficient, when added to his market price, to exceed his salvage price. Thus, as long as his cash income is sufficient to keep him financially solvent, he may soldier on at a very low market price. There may therefore be large numbers of small farmers who would like to move out but, because of very low 'transfer earnings', stay on at very low incomes.[2]

If this is so, there is an argument that government should play a role on grounds of social hardship. If there is an objective of maintaining small farmers in business in certain regions (the hills and uplands, for example) and at an 'acceptable' (to the consciences of politicians) income, government can raise their market price by direct income support – cash handouts to specific categories of farmer. If, on the other hand, the objective is to assist the exodus from agriculture, it can raise their salvage price by re-training younger farmers, pro-

---

[1] G. L. Johnson, 'Supply Function – Some Facts and Notions', in E. O. Heady *et al.*, (eds.), *Agricultural Adjustment Problems in a Growing Economy*, Iowa State UP, Ames, 1958.

[2] This aspect has been investigated and reported on in R. Gasson, 'Occupational Immobility of Small Farmers', *J. Agric. Econ.*, Vol. xx, No. 2, 1969.

viding alternative work in rural areas, assisting with removal and rehabilitation expenses, or by offering special retirement incentives to older farmers (pensions or 'golden handshakes'). From the national point of view it is desirable to assist the transfer of labour from agriculture, which generally has low economic productivity, to industries with higher productivity; and also the transfer of farms from older farmers, often wedded to outdated methods, to younger ones who are likely to be more efficient. This type of structural policy can be justified by market economists as oiling the cogs of the market by reducing the barriers to transfer in response to market forces.

'Voluntary Acceptance of Negative Economic Rent'

The other school of thought on the immobility of the farmer argues that he accepts a low relative income, not because he is forced to by his circumstances, but because he makes a conscious decision to do so. This is certainly so with new entrants into agriculture, and most probably with many of the existing better-educated, larger farmers. We may find that a new university graduate, who has a wide range of employment possibilities and could expect to earn, say, £16,000 a year after a few years as a management entrant to ICI, decides instead to take on a small farm at a high rent with borrowed capital which may yield very little net income (£4,000 a year, say) for a long time. He is apparently willing to sacrifice £12,000 a year to be a farmer. Part of his decision may be explained away by the taxation and financial benefits of farming mentioned earlier (p. 33) – say, £6,000 a year. But, even then, he is still £6,000 a year worse off financially. Yet, in the full knowledge of this financial sacrifice, he still goes into agriculture with all the attendant headaches. What motivates him is a matter for the individual; but many people *are* desperately keen to get into agriculture, almost regardless of the cost, and to remain in it at very low incomes when they have more lucrative alternative opportunities for employment. And it is not only young people who, it might be argued, do not have sufficient experience to know any better who do this. It is also people leaving the armed forces, the civil service and industry in search of a life on the land.

Rural sociologists[1] have explored the motivations and values of

---

[1] For example, R. Gasson, 'Goals and Values of Farmers', *J. Agric. Econ.*, Vol. xxiv, No. 3, 1973.

THE ECONOMIC ARGUMENTS FOR SUPPORT 51

farmers to try to explain this phenomenon. They have come up with answers such as 'intrinsic' orientations (enjoyment of work tasks, preference for a healthy outdoor life, purposeful activity, and independence) and 'expressive' values (pride of ownership, self-respect, meeting a challenge), as well as the social prestige and the 'instrumental' values of expanding a business and safeguarding income for the future. Whatever all these may mean, they are obviously things on which a lot of people place great store and for which they are prepared to sacrifice a lot of money (£6,000 a year in the case of our new graduate). This phenomenon was described by Anne Martin as 'a voluntary acceptance of negative Economic Rent'.[1,2]

If this is the explanation of low relative incomes, there is nothing government should or can do to prevent it. It should do nothing because these farmers are living in a free society and exercising their right to choose their own occupation from a number of available alternatives. It is their business if they choose a low-paid occupation, and their fellow-citizens have no conceivable obligation to attempt to raise farmers' earnings at their expense as taxpayers or food consumers. It can do nothing to prevent it because, if government tries to increase farm product prices or subsidise farm inputs, or even gives direct income subsidies to raise farmers' profits, those profits will, after a time, be eroded by higher rents, land values and interest payments to the point at which the previous differential (the negative economic rent) is restored. The exercise will be a costly futility.

It does not matter whether the asset-fixity theory or the theory of negative economic rent applies in a particular country or region (indeed, it is probable that each will apply simultaneously in most places). Both contain a basic truth, and both have the same implications for the agricultural policies which have been practised in the post-war period. The farm problem is one of supply and demand, as are all economic problems. Fundamentally, however, it is not a problem of the supply of and demand for farm products; it is a problem of the supply of and demand for farmers. It is farmers, not milk and wheat, which are in over-supply. Any agricultural policy which fails

---

[1] A. Martin, *op. cit.*, p. 91.
[2] Economic rent refers to the earnings of any factor of production in excess of the minimum sum necessary to keep it in its existing use. This minimum sum will depend on employment opportunities available to the factor elsewhere.

to recognise this basic truth, and acts on the symptoms rather than the root causes, is doomed to failure. These matters of principle are related to the specific examples of British and European policy in Section V.

Agriculture and the Balance of Payments

This review of the economic arguments for support cannot claim to be exhaustive. Before turning to the arguments *against* support, however, it is worth alluding briefly to a final economic argument which was long used in Britain and has provided additional fertile ground on which agricultural economists could work.[1] This is the question of agriculture's contribution to the balance of payments – or, rather, the increased (beneficial?) contribution to the balance of payments which higher support could bring. It was an argument which the farm lobby pushed very hard in the 1950s and 1960s in their quest for higher support. At a time of recurrent balance-of-payments 'crises' it had considerable superficial appeal – especially in terms of 'net import saving'. This argument ran along the lines that, if British farmers raised their output by an additional £100-worth of produce, it would replace £100-worth of imported food. But since the extra output would require only £20-worth of imported inputs, there would thus be a net import saving of £80.

Such a simplistic proposition, however, neglected a number of important factors which would have to be taken into account to determine the much smaller 'net contribution' to the balance of payments from an agricultural expansion. Amongst other things, we would need to know the opportunity cost in terms of foreign exchange of the additional resources diverted into agriculture for the expansion and of the extra government payments to induce it, plus the reciprocal effects on our exports of a reduction in our food imports.[2] Before deciding whether or not to expand agriculture we would also need to know the relative costs and benefits of stimulating other industries

---

[1] A useful discussion of the issue can be found in G. McCrone, *op. cit.*, Part II.

[2] The first calculation of 'net contribution' was in L. Moore and G. H. Peters, 'Agriculture's Balance of Payments Contribution', *Westminster Bank Review*, August 1965. A debate on this issue between the NFU and an independent economist is contained in A. Winegarten and T. E. Josling, *Agriculture and Import Saving*, Occasional Paper No. 5, Hill Samuel & Co. Ltd., 1970. Also T. Phillips and C. Ritson, 'Agricultural Expansion and the UK Balance of Payments', *National Westminster Bank Quarterly Review*, February 1970.

THE ECONOMIC ARGUMENTS FOR SUPPORT 53

to increase exports or reduce imports. Moreover, other policies to assist the balance of payments – floating exchange rates, devaluation, import controls, and so on – would have to be considered.

In the event, agriculture and other ministers in successive governments paid much lip service to agricultural expansion (as in the National Plan of 1965, the Little Neddy Report of 1968,[1] and the White Paper of 1975[2]) but put little additional hard cash into it after the immediate post-war Expansion Plan. To its credit, the Treasury did not fall for this rather dubious ploy.

Since the UK's entry to the EC, the main balance-of-payments problem has disappeared with the floating of the pound and the advent of North Sea oil. The argument for agricultural expansion is thus nowadays in terms of taking maximum advantage of the generosity of the CAP and other Community policies to minimise our budget contribution, which is a (small) burden on the balance of payments.

---

[1] Economic Development Committee for Agriculture, *Agriculture's Import Saving Role*, HMSO, London, 1968.
[2] *Food from Our Own Resources*, Cmnd. 6020, HMSO, London, 1975.

V. The Case Against Support

In Section IV, the arguments for support of agriculture were examined critically from a theoretical perspective. This Section discusses the effects of support on the farmers, on the national economy, and on the international trading system in the light of British and European experience.

Support and the Farmer

The attitude of the bulk of farmers, as expressed by the NFU and its counterparts, is unquestionably very strongly in favour of government involvement to raise agricultural prices, restrict agricultural imports (whilst encouraging agricultural exports), subsidise the inputs of agriculture, and maintain the farm population. At the same time, they protest at other aspects of state activity which they consider irksome – planning controls on farm buildings, restrictions on farming practices which alter the countryside, the prospect of having to pay rates (property tax) on their land, and so on. They are thus in favour of the ministrations of one set of bureaucrats whilst deploring the attentions of another. This is a position which is becoming increasingly untenable.

While there are very recent signs that this inconsistency is being recognised by farmers, for the most part they cling to the idea that support and the farmer are like Siamese twins: if you kill off one (support), you kill off the other (the farmer). Any criticism of support policy is construed as criticism of farmers as such; critics of support are enemies of the farmer.

This *Hobart Paperback* argues that such notions are incorrect. Support policy as we have known it for the past 25 years is not only unnecessary for the survival of most farmers; it has done most of them little good and has indeed been very harmful to new and potential entrants. Why is this so?

Price Policy

Farmers are pre-occupied above all with farm product prices and have a deeply ingrained idea that they should receive a 'fair' or 'just' price

THE CASE AGAINST SUPPORT 55

for their products regardless of market conditions.[1] The concept of a 'fair price' is a subjective one, but can be taken to mean a price which covers production costs and leaves a 'fair reward' for the labour of the farmer. Farmers are not alone in this view of the world, which implies that farming is for the exclusive benefit of farmers and that consumer requirements can be quietly forgotten. It has, however, no basis in economics.

Nonetheless, governments everywhere have implicitly accepted this view by pursuing policies of price support. Price policy can aim either to stabilise or to raise farm product prices, or to do both. Price stabilisation (but not rigidity) is economically desirable, as was argued in Section IV. But state involvement is not essential to achieve it and is *un*desirable because all recent history shows that it invariably becomes price-raising. Stabilisation implies both raising and lowering prices; farmers will always favour the former and do everything in their power to prevent the latter.

In the long run, however, price-raising becomes self-defeating. It induces even more rapid increases in output in relation to demand than would otherwise occur, and will thus tend to depress market prices still further, according to the terms-of-trade theory already mentioned. This is precisely what has happened in Britain, Western Europe and North America. Price support achieved its goal of increasing output at a time of world food shortage after the Second World War – and far more rapidly than most economists had predicted. By the mid-1950s, eggs, pigs and milk were in over-supply and there was pressure on their prices; by the 1960s virtually all products were in over-supply and the price problem became universal.

At a time when world prices of many products were falling, the government in Britain was faced with guaranteed prices which could be altered only within strict limits under existing legislation.[2] Inevitably, the gap between British and import prices grew larger and had to be bridged by increasingly costly deficiency payments for cereals and fatstock. The world market price of wheat, which stood at £20 per ton in the late 1950s, fell to £17·55 in 1962, whilst the guaranteed price of £28 was reduced by only £1.[3] Consequently, the deficiency

---

[1] This notion is discussed in P. G. James, *Agricultural Policy in Wealthy Countries*, Angus and Robertson, London, 1971, pp. 306-8.

[2] Above, p. 12.   [3] *Source:* Price Review White Papers.

payment per cwt. rose in the year of 1961/2 to 1962/3 from 24 to 55 per cent of the market value. A similar development occurred with fatstock where the unit subsidy rose from 24 to 31 per cent of the market value between 1957/8 and 1961/2. Every other country which supported the prices of its farm products had a comparable experience. It has been estimated[1] that aggregate price support (overt and covert) rose in the period 1956 to 1966 from 18 to 47 per cent of market values in France; from 22 to 54 per cent in West Germany; from 27 to 54 per cent in Sweden; and from 16 to 51 per cent in the original six-member EEC as a whole.

Under these circumstances most governments felt they could no longer afford to go on raising guaranteed prices in line with prices generally; the burden was becoming too heavy. The Conservative Government of the early-1960s managed to stabilise farm product prices by the stealthy expedient of the standard quantity – which apparently maintained the guaranteed price announced at the Annual Review but reduced it retrospectively if farmers collectively exceeded the standard quantity (that is, the output deemed desirable by government).[2] This was as unfair as offering a carrot to a donkey and then beating him for taking it! It also destroyed one of the main arguments for a guaranteed price: that it is known in advance and therefore eliminates one of the main uncertainties the farmer faces. Nevertheless, along with minimum import prices for cereals,[3] it did enable guaranteed prices and support costs to be kept under control – to the chagrin of farmers who, as a result, had by the late-1960s become much more favourably disposed than previously to the idea of joining the EC.

The trend of farm prices in the UK is illustrated in Table VI in which the MAFF price index (including subsidies) of all agricultural products has been converted to a common base period (1936-38=100)

---

[1] R. W. Howarth, *Agricultural Support in Western Europe*, Research Monograph 25, IEA, London, 1971 (figures calculated from Table 3, p. 29).

[2] Standard quantities, or similar arrangements under different names, were introduced for: pigs (1961), eggs (1963), cattle and sheep (1964), and wheat and barley (1964). There had been a standard quantity for milk since de-control in 1954, and sugar-beet output was limited from 1951 under the Commonwealth Sugar Agreements. The Potato Marketing Board (1955) attempted to restrict output by acreage quotas. The Minister of Agriculture, then Mr Christopher Soames, found a general 'let out' from the hitherto largely open-ended guarantee arrangements in the phrase in the preamble to the Agriculture Act, 1947, which refers to '... producing such part of the nation's food ... as in the national interest it is desirable to produce in the United Kingdom, ...'.

[3] Above, pp. 12-13.

and then adjusted for inflation according to the Retail Price Index to give an index of real agricultural prices over the period from 1935 to 1981. In money terms, farm product prices almost doubled during the war, rose by a further 47 per cent during the post-war expansion (1946-52), increased slightly (by 7 per cent) up to the mid-1950s, and then, as surpluses emerged and support costs rose, held virtually constant for 11 years (from 1956 to 1966 inclusive), during which time inflation eroded all the war-time and immediate post-war gains. By 1960, real agricultural prices were back to where they had been before the war.

Money prices began to move upwards once more in the late-1960s and took off like a rocket in the early 1970s, rising by 150 per cent in the period 1972-77, during the commodity boom. At that stage there were widespread fears of a long-run shortage of temperate foodstuffs and farmers hoped that the post-war decline in the terms of trade against agriculture would be reversed. The rise in the world market prices of some agricultural products was such that, for a short time in 1973-74, cereal and sugar prices on the world market were considerably higher than their support prices in the EC. The import levies became export taxes and the Commission began to think that the CAP might not be such a bad thing after all,[1] having been highly critical a few years earlier in the Mansholt Report of its own creation. But the commodity boom turned out to be a freak combination of circumstances[2] – which were short-lived. British farmers did make up much of the lost ground on real prices, which had been eroded to 82 per cent of their pre-war value by 1971, but only for the four years of 1974 to 1977 (during the high point of which they stood at no more than 9 per cent above their real value in 1936-38). Since then, a 33 per cent money increase (1977-81) has been completely wiped out by inflation and by 1981 farmers' real prices were back to their 1971 levels. Entry into the EEC, which was meant to herald a

---

[1] EC Press and Information, 'The CAP Serves Farmers and Consumers in a Time of Economic Instability', *Newsletter on the CAP*, June 1975.

[2] The prices of grain, butter and sugar all increased considerably during the early-1970s. There was an unprecedented rise in the price of butter in 1971 as the result of poor weather in Europe during the summer of 1971, and in Australia and New Zealand in both 1970-71 and 1971-72, which reduced milk output. The world price of grain rose dramatically in 1972-73, mainly due to large purchases by Russia and China. Sugar prices, always volatile, shot up to £460 a ton by October 1974 at a time when the Commonwealth Sugar Agreement price was £80 and that of the EEC £118.

TABLE VI
UK INDICES OF THE PRICES OF ALL AGRICULTURAL PRODUCTS, 1935 TO 1981

(Based on prices received by farmers after addition of subsidies where payable)

Year	Index of Prices of All Agric. Products, 1936-38=100			Purchasing Power of the £	Real Index of Agric. Prices
1935	93			104	97
1939	103 ⎫			95	98
1940	142 ⎪			79	112
1941	171 ⎪	War-time		70	120
1942	183 ⎬	+91.3%		65	119
1943	185 ⎪			62	115
1944	192 ⎪			61	117
1945	197 ⎭			60	118
1946	208 ⎫	Post-war		58	121
1947	241 ⎬	Expansion	Selective	54	130
1952	306 ⎭	+47%	Expansion	41	126
1956	328 ⎫		(1952-56)	36	118
1957	319 ⎬		+7%	35	112
1958	327 ⎪			34	111
1959	317 ⎪			34	108
1960	302 ⎪			33	100
1961	314 ⎬	The Lean		32	100
1962	313 ⎪	Years		31	97
1963	309 ⎪	Almost no		31	96
1964	317 ⎪	change		30	95
1965	321 ⎪			29	93
1966	330 ⎭			27	89
1967	333			26	87
1968	342			25	86
1969	355			24	85
1970	375			22	83
1971	391			21	82
1972	414 ⎫			20	83
1973	551 ⎪	Commodity		18	99
1974	622 ⎬	Boom		16	100
1975	771 ⎪	+150%		13	100
1976	992 ⎪			11	109
1977	1,031 ⎭			10	103
1978	1,062			9	96
1979	1,173			8	94
1980	1,239			7	87
1981	1,372			6	82

THE CASE AGAINST SUPPORT

SOURCES OF TABLE OF AGRICULTURAL PRICE INDICES, 1935-1981

1. Purchasing Power of the Pound
    1935-77 from *The Economist Diary 1979*, p. 24.
    1978-81 Calculated from RPI figures given in *Annual Review of Agriculture, 1982*.
    p. 12, and *1984*, p. 15.

2. Agricultural Price Indices
    1935-46 *Annual Abstract of Statistics, 1935-46*, p. 255.
    1947   *Annual Abstract of Statistics, 1954*, p. 282.
    1952-60 *Annual Abstract of Statistics, 1961*, p. 311.
           All these give as a base 1936-38=100
    1961-68 *Annual Abstract of Statistics, 1969*, p. 362.
           Base 1954/55 – 1956/57=100 (=322)
    1969-71 *Annual Abstract of Statistics, 1973*, p. 390.
           Base 1964/65 – 1966/67=100 (=323)
    1972   *Annual Review of Agriculture 1977*, p. 12.
           Base 1968/69 – 1971/72=100 (=366)
    1973-76 *Annual Review of Agriculture 1978*, p. 12.
           Base 1970=100 (=375).
    1977-81 *Annual Abstract of Statistics, 1983*, p. 344.
           Base 1975=100 (=771)

---

major price boost for British farmers, had done nothing at all for them in real terms. A tonne of wheat, for which they received £9·90 in 1938, brought in £108·92 in 1981; but to have maintained the 1938 purchasing power, £165 would have been required – that is, over 50 per cent more than the 1981 price. A litre of milk, at 1·2p in 1938 and 13·79p in 1981, would have had to fetch 20p in 1981 to maintain its value – 45 per cent more than it did. Fat cattle, however, fetched 4·6 pence per kg. liveweight in 1938 and 88·73p in 1981, thus *more* than maintaining their real price which had increased 19-fold whilst prices generally had risen 16·6 times.[1] Cereals and milk were in far larger surplus than beef in 1981.

As Britain showed in the 1960s, governments may be willing to spend a large slice of taxpayers' money on guaranteeing prices for farmers, but the total amount is finite and, at some stage depending upon the political situation, a halt will be called to the rising cost. In the European Community it has taken much longer to reach that limit – if, indeed, it *has* now been reached – but the pressures are the same. In the Community as a whole the target prices, expressed in national currencies, rose on average by 62 per cent in money terms in the period from 1972-73 to 1976-77, and by 40 per cent between 1976-77 and 1980-81. In *real* terms, however, they increased by only 8 per cent in the first period, and actually declined by 8 per cent in

---

[1] *Sources:* calculated from *Annual Abstract of Statistics* and Price Review White Papers.

the second.[1] The averages conceal variations between countries due to the manipulation of 'green' currencies; but the decline in real prices in the later period occurred in all member countries varying from 11 per cent in Italy to 3 per cent in Denmark and Ireland, the UK figure being 9 per cent. Even the farmers of France and Germany, with their well-known political clout, have been unable to persuade their governments to insulate them from the long-run decline in the terms of trade of agriculture. McCrone's theory has remained valid after two decades.

Sooner or later, despite all the efforts of governments at the expense of taxpayers and consumers, the forces of supply and demand will exert their crucial influence because the cost of trying to frustrate them will become insupportable. And other problems will arise in the course of such efforts. The signals which prices send between consumers and producers will be muffled if not completely stifled. The general level of agricultural prices and the relative prices of different agricultural products will be determined not by supply and demand but by political horse-trading. For price policy is not simply a matter of politicians and civil servants raising all farm prices in Britain or the EC by the same amount above their market levels; the prices of different products will be increased by different amounts, largely in response to the political muscle of different producer groups. Those who are the most numerous (milk producers) and the most influential (cereal growers) will tend to benefit at the expense of the others (beef, pigmeat, and poultry), causing resentment and tension between groups. The resentment will be greatest (as it has been, for example, with British pig producers since EC entry in 1973) when price policy is of the type which artificially raises the internal market price of one group (cereals) by support buying and import levies and thereby increases the input costs of another group to such an extent that they exceed the benefit the latter group derives from support of its final product (pigmeat). Even though the state may not intend it, the 'effective protection' of some farmers may therefore be negative; the net result of policy is to make them worse off than with no policy at all. The relative disadvantage of livestock producers is still not recognised by the EC. By 1983 (provisional figures), British crop prices had risen by 35 per cent since 1980 and by 204 per cent

---

[1] *Source*: M. Tracy, *op. cit.*, pp. 331-2.

since EC entry, whilst the prices of livestock for slaughter had increased by only 22 per cent and 188 per cent respectively (all in money terms).[1] In the period from 1977-78 to 1983-84 (forecast), the real net farm incomes of pig and poultry farmers will have fallen by 75 per cent whilst those of cereal farmers will have increased by 60 per cent.[2] Because of its political nature, price support will inevitably be inequitable between producers of different products.

Inequitable Distribution

Moreover, price support is inevitably inequitable between farmers of different sizes. Since it is paid per unit of output, those who produce the largest output will obviously receive the most support in absolute terms. Those who produce 2,000 tons of wheat will receive 10 times as much as those who produce 200 tons. The same applies to non-specific production grants (input subsidies) because the larger farmers also use more inputs. It is therefore likely that the bigger farmers, who are the richer ones, will gain most from the total support system. As far back as 1957, E. M. H. Lloyd estimated that something like two-thirds of all price subsidies went to one-third of the full-time (larger) farmers.[3]

More sophisticated studies have been carried out since then. Josling and Hamway's work,[4] relating to 1969, divided British farmers into four groups (quartiles) according to their incomes. The first quartile represented the poorest 25 per cent of farmers, and the fourth the richest 25 per cent. The support scheme for potatoes (operated by the Potato Marketing Board) and the deficiency payments scheme for cereals handed 80 and 66 per cent respectively of the support into the pockets of the richest quarter of producers and only 9 and 7 per cent to the poorest quarter; the two intermediate quartiles together received 11 and 27 per cent. The richest quartile took 58 per cent of fertiliser and lime subsidies; the poorest received 10 per cent. Only in the support systems for cattle and sheep did a fairly large proportion

---

[1] Cmnd. 9137, op. cit., p. 15.
[2] Ibid., p. 44. For a detailed discussion of effective protection, J. Strak, *Measurement of Agricultural Protection*, Macmillan, London, 1982.
[3] E. M. H. Lloyd, 'Some Thoughts on Agricultural Policy', *J. Agric. Econ.*, February 1957.
[4] T. E. Josling et. al., *Burdens and Benefits of Farm Support Policies*, Trade Policy Research Centre. London, 1972, pp. 50-82.

of the payments go to the poorest quartile (18 and 21 per cent respectively), which was presumed to reflect the importance of cattle and sheep in areas where farming incomes are low. Taking price and input subsidies together, it was reckoned that the better-off farmers who earned 50 per cent of the total income were receiving 71 per cent of the total payments, whilst the poorest half received only 29 per cent. This result was predictable. What was surprising, however, was that, overall, the richer farmers were receiving the most, not only absolutely but also relatively to their income shares without support. The richest quartile, which would have accounted for 41 per cent of total non-supported income, received 43 per cent of total support payments and 48 per cent of the total income. This outcome was largely a reflection of the concentration of support payments on cereals and cattle which were produced mainly by the larger and richer farmers. The former British policy was therefore regressive, extremely wasteful, and inequitable in handing out the most support to the richest farmers and the least support to the poorest. Josling and Hamway predicted that full acceptance of the CAP, with its even stronger emphasis on cereals, would increase the inequity and the waste by handing a further 2 per cent of the income (raising it to 50 per cent) to the richest quartile.

Since Josling and Hamway's 1972 study, no-one has attempted such a detailed estimation of the distribution of the benefits of farm support in the EC, but the likelihood is that the regressiveness and inequity have persisted, and indeed increased, because the policy has remained essentially one of blanket price support with even less emphasis on specific production grants than the former British system and even more emphasis on cereal and milk prices.

Real and Relative Income Largely Unaffected

That support cannot maintain real prices for agricultural products has already been demonstrated. But what of farm *incomes*? To obtain a true picture, it is necessary to examine data over a considerable period (1937/38 to 1983) and to adjust them for inflation. Net Farm Income (for which there are published figures up to 1978 in the Price Review White Papers) and its successor, 'Farming Income' (which was introduced in 1980 but for which there are now published figures going back to 1970), are the only statistics which are readily available

THE CASE AGAINST SUPPORT

and which we have to use, regardless of their limitations (discussed in Section IV). 'Farming Income' tends to give a higher figure than NFI because it does not deduct rent and treats the national farm as owner-occupied as opposed to tenanted.

It can be seen from the figures given in Table VIIA that aggregate money income has increased spectacularly over the period from 1937/38 to 1983, but that real income has hardly progressed at all. Using the NFI figures up to 1976 (those from 1970 onwards are not given in the Table but are taken from Price Review White Papers) and taking three-year moving averages to iron out the annual fluctuations, it can be seen that these were substantial gains in *real* income (as in prices) during the Second World War and the post-war expansion – income almost doubled during the war and had almost trebled by the end of the expansion – but that it began to even out during the 1960s and had fallen by the early 1970s. The fall between '1961' (1960-62 average) and '1971' (1970-72 average) was 14 per cent.[1] Even during the height of the commodity boom ('1971-76'), real NFI increased by only 9·5 per cent. Between '1961' and '1976' there was an overall fall of 5·9 per cent.

Using total real farming income as a measure (Table VIIB), figures are available in the 1984 White Paper for the period from 1970 (all at constant 1980 prices) – which covers the period the UK has been in the Common Market (1973 onwards). Real farming income reached a one-year peak in 1973, which has not been attained since then. In fact, it recorded a considerable decline of 41 per cent over the decade from '1971' to '1981' and a decline of 43 per cent from the peak in '1973' to '1981'. Even during the period of the commodity boom, this measure actually declined by 4·4 per cent.

Since, however, the aggregate income has been shared by fewer farmers over the years, we should also examine the trend of income per farmer. Surprisingly, despite well over a century of agricultural statistics, the Ministry of Agriculture did not publish data on the number of farmers until quite recently: the first year available is 1970. For earlier years we have to make rough estimates based on the number of holdings. A reasonable measure of the number of full-time farmers is the number of holdings of 15 acres or more. Using that

---

[1] Henceforth, years inside quotation marks connote a three-year period comprising the year mentioned plus the immediately preceding and following years, over which some variable is expressed as an annual average.

TABLE VIIA

AGGREGATE NET FARM INCOME AND FARMING INCOME IN MONEY TERMS AND IN REAL TERMS AND PER FULL-TIME FARMER, 1937/38 TO 1983

Year	£m Money Aggregate NFI	£m Real Aggregate NFI	Number of F/T Farmers '000	£ Real NFI/ F/T Farmer
1937/38	56·5	56·5		
1946/47	194·0	112·5		
1952/53	347·5	142·5		
1956/57	338	122	306	399
1959/60	362	123	295	417
1960/61	389·5	129	293	440
1961	424	136	289	471
1962	445	138	279	495
1963	406	126	275	458
1964	436	131	273	480
1965	412	119	269	489
1966	457	123	265	464
1967	509	132	260	508
1968	448	112	274	409
1969	498	119	264	451
	(Farming Income)	(Farming Income)		(Farming Income)
1970	567	125	216	579
1971	640	134	230	583
1972	682	136	229	594
1973	952	171	222	770
1974	803	128	214	598
1975	1,005	131	212	618
1976	1,293	142	219	648
1977	1,269	127	212	599
1978	1,252	113	216	523
1979	1,141	91	215	423
1980	1,018	71	208	341
1981	1,318	79	204	387
1982	1,802	104	203	512
1983 (forecast)	1,536	85	201	423

TABLE VIIB

REAL FARMING INCOME PER FULL-TIME FARMER

(*at constant 1980 prices*)

Year	Total Real Farming Income £m.	Per F/T Farmer £
1970	2,046	9,472
1971	2,107	9,161
1972	2,097	9,157
1973	2,677	12,058
1974	1,944	9,084
1975	1,964	9,264
1976	2,168	9,899
1977	1,843	8,693
1978	1,680	7,778
1979	1,344	6,251
1980	1,018	4,894
1981	1,181	5,789
1982	1,486	7,320
1983 (forecast)	1,211	6,025

SOURCES OF TABLES VIIA AND VIIB

Aggregate NFI 1937/38 to 1969
Based on the 'old' definition of NFI – the reward for manual labour, management, and return on tenants' capital. From 1964 NFI excludes stock appreciation. Figures from *Annual Review and Determination of Guarantees*, White Papers always taking the figure for a given year from the last White Paper which gives it, as there have been major revisions to some figures after two or three years from their first publication.

Number of Full-time Farmers
Up to and including 1969 the number of F/T farmers is a 'guesstimate' based on the number of holdings of 15 acres and over in size. Figures from MAFF, Agricultural Statistics, various. From 1970 figures for the number of F/T farmers have been given in the *Annual Review*, White Papers; again the last figure published for a given year has been used.

Purchasing Power of the Pound 1936-38 = 100.
Sources as in Table VI.

Farming Income in Money and Real Terms 1970-83
Money Farming Income: actual figures from *Annual Review of Agriculture 1984*; Real Farming Income calculated from index published in the same source (1980 = 100) – both from p. 40.

measure, the extreme right-hand column of Table VIIA gives a rough picture of real NFI per full-time farmer up to 1970. Over the period from '1956/57' to '1968' it increased by 14 per cent. Between '1960' and '1968' it increased by about 3 per cent. Farmers in Britain had declined in numbers and increased their productivity sufficiently to raise their real incomes slightly despite a falling *aggregate* real income.

Since 1970, the position has changed. From the more accurate figures for the number of farmers and the farming income measure (Table VIIB), it is not possible to compare directly the situation before and after 1970, but there is no doubt that real income per farmer increased very rapidly in the early-1970s (boosted by the freak year of 1973), though it began to tail off in 1978 and fell to a dramatic low in 1980 when, at £4,894, it was only 41 per cent of the 1973 peak and 61 per cent of the 1970 figure. Taking the three-year averages for '1971' and '1981' (£9,263 and £6,001), there was a drop in real income per farmer of 35 per cent over the decade.

It is possible to infer from Tables VIIA and VIIB that the relative income position of British agriculture must have declined over the years. This is illustrated in Table VIII. Relative income, which had been as high as 120 per cent in 1950 (Table IB), fell to around parity in 1955-56. It fell further during the 1960s to 88 per cent in 1967-69, recovered during the early-1970s to regain parity in 1976, but has since dropped back to 82 per cent (1980-82).

Only when market forces have been moving in the farmer's favour has his real income increased. As was seen above, for most of the period they have been moving against him, and real income from the national farm has declined such that, in the last three years (1981-83), it has been on average 37 per cent less than its post-war peak in 1952-53. The recovery during the commodity boom was short-lived. The expenditures of taxpayers and consumers (to be discussed in the next section) have been unable to halt the decline in the total income of agriculture and, although they have *maintained* real income per farmer, it has not increased rapidly enough to keep pace with incomes in general. The 'farm problem' persists. Even the mighty resources of the European Community have not been able to overcome the constraints on farm income imposed by the characteristics of the demand for food.

Furthermore, neither the old British policy nor the CAP has been able to achieve its declared objective of stabilising income, as a glance

TABLE VIII

RELATIVE INCOME OF AGRICULTURE: UK, 1955-56 TO 1982

	Agricultural manpower as a percentage of total civilian manpower (a)	The contribution of agriculture to GDP (b)	Relative income[1]
1955-56	4·8	4·7	98
1963-64	3·9	3·6	92
1967-69	3·3	2·9	88
1973	2·9	2·8	96
1974	2·7	2·5	93
1975	2·7	2·6	96
1976	2·7	2·7	100
1977	2·7	2·5	93
1978	2·7	2·4	89
1979	2·6	2·2	85
1980	2·6	2·1	81
1981	2·7	2·2	81
1982	2·7	2·3	85

[1] Relative income = $\frac{b}{a} \times 100$.

Sources: 1955-56 and 1963-64 from Table II.
1967-69 onwards: Annual Review White Papers 1978, 1980 and 1984.

at the right-hand column of Table VIIA shows. Table IX (p. 68) reveals that, over the 10 years from 1959-60 to 1968-69, real income per farmer increased on six occasions and fell on four, from one year to the next. The average annual variation was 7·6 per cent, with actual variations ranging from −19.5 per cent (1967-68) to +1·8 per cent (1964-65). In the 13 years from 1970-71 to 1982-83, the annual variations have increased considerably. Income per farmer rose on seven occasions and declined on six, the average annual variation being 14·2 per cent. This average conceals some very large variations (+29·6 per cent in 1972-73; −22·3 per cent in 1973-74; +32·3 per cent in 1981-82) as well as some very small ones (+0·7 per cent in 1970-71 and +1·9 per cent in 1971-72).

The Cost of Support

Support may not have done farmers in general much good, but it has certainly cost a great deal over the years. To quantify the full

TABLE IX
ANNUAL PERCENTAGE VARIATIONS IN REAL INCOME PER
FULL-TIME FARMER, 1959-60 TO 1982-83
(*Derived from Table VIIA*)

Year	Annual Variation %
1959-60	+5·5
1960-61	−7·0
1961-62	+5·1
1962-63	−7·5
1963-64	+4·8
1964-65	+1·8
1965-66	−5·1
1966-67	+9·5
1967-68	−19·5
1968-69	+10·3

Annual Average Fluctuation 7·6%

1970-71	+0·7
1971-72	+1·9
1972-73	+29·6
1973-74	−22·3
1974-75	+3·3
1975-76	+4·8
1976-77	−7·6
1977-78	+12·7
1978-79	−19·1
1979-80	−19·4
1980-81	+13·5
1981-82	+32·3
1982-83	−17·4

Annual Average Fluctuation 14·2%

cost is to enter a statistical and economic minefield. There is, however, no disputing the budgetary cost, that is, the amount contributed by taxpayers.

The figures in Table XA show that, during the entire period from the introduction of deficiency payments in 1954-55 to the present, the cost to the Exchequer of agricultural support, as defined by MAFF, has totalled £13,579·6 million, of which £5,022·6 million was spent before Britain joined the EC in 1973, and the remaining £8,557 million since then. These figures do not include the cost of research, education, and extension (except for the first three years, which include research and advice). Nor do they include the cost of the various other activities of MAFF, such as subsidies to forestry and fishing. As the proponents of the support system[1] point out, the Exchequer cost has not increased in real terms over the years (Table XB) – though it has risen sharply in the past few years towards the '1956' figure – and has decreased considerably as a percentage of total government expenditure (which is a result not so much of the decline in agricultural expenditure as the increase in total government expenditure). And, of course, the cost is small relatively to health and welfare spending. Nevertheless, agriculture does consume considerable sums of public money – equivalent at present to almost double the UK's overseas aid programme, 57 per cent more than is spent on law and order, a comparable sum to that spent on transport, and one-third of total expenditure on all other industries, trade, energy, and employment together.[2]

But budgetary cost is misleading as an indicator of the full cost of support. Under the former British system, where most products (except milk) were supported by deficiency payments, it gave a good guide. But with the advent of the CAP régime it tells only a fraction of the full story because the variable import levy/intervention buying system artificially raises the internal market prices of most products, thereby imposing a considerable burden on the consumer of food as well as on the taxpayer.

The only way of estimating the cost of a wide variety of simultaneous policies for product prices is to try to discover the difference between what is paid for those products under the support system and what would be paid without it: in other words, to re-value domestic agricultural output at 'import parity prices'. This was a method pioneered by the late Professor Eric Nash,[3] later used by Gavin

---

[1] For example, an article by Michael Murphy in *The Farmers' Club Journal*, No. 67, February/March 1984.
[2] These figures refer to 1983-84 (estimated outturn) from: HM Treasury, *The Government's Expenditure Plans 1984-85 to 1986-87*, Cmnd. 9143, HMSO, London, 1984, Vol. I, p. 12, Vol. II, p. 13.
[3] E. F. Nash, *Agricultural Policy in Britain*, University of Wales Press, Cardiff, 1965, Ch. V (which is a reprint of the original 1955 article).

TABLE XA
EXCHEQUER SUPPORT TO AGRICULTURE, 1954/55 TO 1983
(at current prices)

Year	£ million
1954/55	197·3
1955	208·4
1956	242·4
1957	288·0
1958	241·4
1959	256·9
1960	263·4
1961	343·2
1962	310·2
1963	294·5
1964	265·1
1965	237·6
1966	229·1
1967	261·5
1968	269·1
1969	268·3
1970	256·5
1971	322·7
1972	267·0
Sub-total	5,022·6
1973	392·0
1974	494·5
1975	511·6
1976	378·4
1977	460·1
1978	536·9
1979	677·0
1980	1,012·4
1981	972·4
1982	1,432·8
1983	1,688·9
Sub-total	8,557·0
Total	13,579·6

SOURCES FOR TABLE XA
Total cost of support:
1954/55 – 1957/58: McCrone, op. cit., p. 46.
1958/59 – 1959/60: Annual Review White Paper 1967.
1960/61 – 1967/68: Annual Review White Paper 1969.
1968/69 – 1983/84: Annual Review White Papers 1973-1984, always using the last White Paper in which a figure for a particular year is given.

TABLE XB

EXCHEQUER SUPPORT TO AGRICULTURE AT CURRENT VALUES, AT 1981 VALUES, AND AS A PERCENTAGE OF TOTAL GOVERNMENT EXPENDITURE, '1956' TO '1981'

	Exchequer Support for Agriculture (3 year average) (at current prices) £ million	(at 1981 prices) £ million	Total Government Expenditure (at current prices) £ million	Support as percentage of Total Government Expenditure %
'1956'	246·3	1,477·8	4,868	5·0
'1966'	242·7	1,092·1	9,541	2·5
'1970'	282·5	1,017·0	13,526	2·1
'1976'	450·0	810·0	39,372	1·1
'1981'	1,289·2	1,289·2	84,785	1·5

Sources: Exchequer support: as in Table XA.
Total government expenditure: from Annual Abstract of Statistics 1983, p. 284, and previous issues.

McCrone,[1] and developed by the present author in a previous IEA publication[2] which discussed the problems and limitations of this method in more detail than is necessary here. The main difficulties are in the selection of suitable import prices and in the implicit assumption that the supply of agricultural imports is perfectly elastic at current import prices – that is, that if domestic prices were reduced to import parity ones, we could make up whatever domestic supply was lost by increasing imports at current import prices. That assumption is obviously questionable, but it is difficult to find a better one since we do not know the price elasticities of all the different agricultural imports into Britain, let alone into other countries if we want to make international comparisons. Nor do we know how British agriculture or other supported agricultures would adjust to

---

[1] G. McCrone, op. cit., Ch. 2.   [2] R. W. Howarth, op. cit.

import parity prices. The adoption of the assumption must lead to some over-estimation of the amount of support, but the method is the best indicator available.[1] It is also very useful in estimating the *relative* costs of support between countries.

Critics of the method who argue that there are no such things as import parity prices, or that they are meaningless, and that it is therefore pointless to carry out the exercise, are usually people who have a vested interest in the public *not* knowing the cost of support. The researcher must, however, be careful to choose import parity prices which reflect the prices of products offered by substantial exporters who are not themselves heavily supported. The studies of the 1950s and 1960s by Nash and McCrone, for example, used Argentinian prices for cereals and beef and Danish prices for pigmeat, eggs and milk. In the 1970s and 1980s it has become more difficult to find realistic import parity prices.

The results of four different estimates of the full cost to consumers and taxpayers during the period 1956 to 1980 are given in Table XI. Naturally, there are some differences in methodology, assumptions, and sources between the estimates, and they have to be subjected to various qualifications. They do, however, bring out clearly the large gap between the Exchequer costs of support and the *full* costs of support, which have been borne by the consumers to a much larger extent under the CAP than before, and the rising trend of support since EC entry. The real burden of support (at 1980 prices) can be seen to have virtually doubled since the 1950s and 1960s (when there was little change in Britain) to well in excess of £3,000 million in 1980, a figure on which Morris[2] and Buckwell[3] are in remarkably close agreement. In 1980, consumers were paying around two-thirds of the total cost compared with less than one-half under the old British system.

The net cost to the United Kingdom of the CAP is arrived at by deducting the benefits received by farmers ('producer gain') from the

---

[1] C. N. Morris, 'The Common Agricultural Policy', *Fiscal Studies*, IFS, London, March 1980. In this article Morris, after considering a number of possible alternative assumptions, concludes that '... in many cases world prices would not be very different from what they are now' if EC countries were free to follow independent farm policies.

[2] *Ibid.*

[3] A. E. Buckwell *et. al.*, *The Costs of the Common Agricultural Policy*, Croom Helm, London, 1982.

TABLE XI

SOME ESTIMATES OF UK TOTAL SUPPORT COSTS, 1956 TO 1980

£ million

	(1) Price Support (estimated)	(2) Consumer Costs	(3) Production Grants & Subsidies	(4) Taxpayer Costs	Total Cost (1)+(3) or (2)+(4)	
					At current prices	At 1980 prices
1956	308	166	58	200	366	1,867
'1966'	320	199	118	239	438	1,664
1978	n.a.	1,787	n.a.	981	2,768	3,598
1980	n.a.	2,303	n.a.	1,252	3,555	3,555

Sources: 1956: McCrone, op. cit., pp. 46, 51, 182.
'1966': Howarth, op. cit., pp. 29, 33. '1966'=average of 1965, 1966, 1967.
For both 1956 and '1966', consumer costs have been estimated by deducting the taxpayer-financed price guarantees from estimated price support. Milk accounted for the bulk of consumer costs in those days.
1978: C. N. Morris, 'The Common Agricultural Policy', Fiscal Studies, IFS, London, March 1980.
1980: A. E. Buckwell, et. al., The Costs of the Common Agricultural Policy, Croom Helm, London, 1982. The figures used are those quoted by D. R. Colman in CEAS/CAS, Agriculture: the Triumph and the Shame: an Independent Assessment, Wye College and Reading University, June 1983.

The figures of Morris and Buckwell refer only to the costs of the CAP. To indicate the full cost of agricultural policy (CAP+national policies), the Exchequer cost of British policy has been added to their figures of taxpayer costs. These additions are items II, III, and IV in the tables on 'Public expenditure under the CAP and on national grants and subsidies' in the 1983 and 1984 Annual Review White Papers (Cmnd. 8804 and Cmnd. 9317). For 1978, the average of 1978-79 and 1979-80 has had to be taken because some payments due in 1978-79 were delayed until 1979-80 by industrial action.

Notes: (i) Price Support (estimated) and Production Grants and Subsidies are not available as such for 1978 and 1980 but are included within Consumer Costs and Taxpayer Costs for those years.

(ii) All the figures for Total Cost represent the burden of farm support shouldered by taxpayers and consumers. The figures for 1956 and '1966' (prior to entry into the EC) were the cost of supporting British agriculture only. The figures for 1978 and 1980 represent the cost of supporting British farmers plus the additional cost to British consumers of paying higher prices for their imported food as a result of CAP price regimes plus the cost to British taxpayers of their net budgetary contribution to the CAP.

total cost of the CAP as such (excluding national policy costs). Morris produces a figure of £1,370 million for 1978 and Buckwell one of £1,651 million for 1980. These net losses are either transferred to other farmers in Europe in the form of higher prices for the food products which the UK purchases from them or are swallowed up in the complex mechanism of the CAP. The burdens of the support system on the general public are considerable. Morris's estimate of the loss to the consumer and taxpayer from the CAP alone works out at £41 per head of population. Taking the Buckwell estimate for 1980 plus the UK national Exchequer support cost, the figure rises to £63·50 per head – an aggregate sum equivalent to just under 1·5 per cent of GDP.

Richard Body[1] also produces calculations of the overall cost of support (which he attributes to the Institute of Fiscal Studies (IFS)). Unfortunately, his figures are presented without explanation and the term 'overall taxpayers' support', used in the title of Table II on page 9 of his book and as a heading to one of its columns, is incorrect or a misprint, as Professor David Colman[2] has observed. For example, the figure Body gives of £1,329 million for 'overall taxpayers' support' in 1978-79 is obviously meant to represent overall support from both taxpayers and consumers. Yet it does not tally with the other IFS figure of £3,191 million for the total cost of the CAP to Britain,[3] nor with my figure of £3,555 million when the cost to the taxpayer of British national policies is added. Body's figures here remain unclear.

Body also lists some taxation advantages which he says are enjoyed by farmers as additional costs of support and which he estimates add a further £1,000 million a year to the total cost. These are 'first-year allowances' on capital equipment purchases; exemption from VAT; special treatment under the Capital Transfer Tax (CTT); and relief from rates (property tax). It is doubtful whether these items should be included. Capital allowances, which have since been considerably changed in the 1984 Budget, are common to all businesses – agriculture does not receive special treatment in this respect. Agriculture is not strictly 'exempt' from VAT; it is 'zero-rated', that is, it does not

---

[1] R. Body, op. cit.
[2] D. R. Colman, 'The Free Trade Alternative', in CEAS/CAS, op. cit., p. 48.
[3] C. N. Morris, op. cit.

pay VAT on its sales but can reclaim the VAT charged on its taxable inputs. It thus receives the same VAT treatment as new building, fuel and power, and several other activities. Agriculture does have some special CTT privileges but they normally benefit, as Body concedes, only the slightly more than one-half of farmers who are also landowners. In any event, it is open to other businessmen, as well as to farmers, to take advantage of various schemes for mitigating or completely avoiding the impact of CTT. Relief from rates *is* a special benefit to farmers. But, as Colman argues,[1] it is not a barrier to free trade and the imposition of rating would not affect farm output or income substantially; it would, rather, lead to an equivalent reduction of rents and land values. Nevertheless, the exclusion of farm land from rating represents a loss of revenue, mainly to the shire counties, of around £400 million a year, according to estimates by *The Economist*.[2] Body is correct to add in the further loss to the Treasury from the reduced excise duty on agricultural fuels; but the sum of £3 million a year, which he suggests, is insignificant in relation both to total farm support and total government revenues.

Distribution of the Burden of Support

It is not only the substantial costs imposed by the support system on every man, woman and child that should be a cause for concern, but even more so that the burden is distributed extremely unfairly among them. The inequitable distribution of the benefits among farmers has already been mentioned; this inequity also applies to consumers. Josling and Hamway[3] had emphasised in 1972 that adoption of the CAP would impose highly regressive support costs compared with the old deficiency payments system which was mainly progressive in incidence because more than half of the support costs were financed by general taxation, of which one-third or so was raised from (progressive) income tax and surtax.

Table XII reproduces the figures Josling and Hamway gave on the incidence of the costs of the two different systems of support in relation to the final income shares of households (income after taxes and state

---

[1] D. R. Colman, *op. cit.*
[2] 'Think Tank: Reducing Agricultural Subsidies', *The Economist*, 26 November 1983, p. 30.
[3] Josling *et al.*, *op. cit.*

TABLE XII

PERCENTAGE SHARES OF SUPPORT COSTS UNDER DEFICIENCY PAYMENTS SYSTEM AND CAP BY HOUSEHOLD INCOME QUARTILE COMPARED WITH FINAL INCOME* SHARES (1969)

	First Quartile (Poorest)	Second Quartile	Third Quartile	Fourth Quartile (Richest)
Share of Final Income	11	19	27	43
Share of Farm Support Costs:				
(1) Deficiency Payments System	7	19	27	47
(2) CAP	13	25	30	32

*Final income is the income remaining after the impacts of all taxes, both direct and indirect, and all government benefits, have been apportioned.
Source: T. E. Josling et al., Burdens and Benefits of Farm Support Policies, Trade Policy Research Centre, London, 1972, p. 77.

benefits have been apportioned). The richest quartile of households had 43 per cent of final income, bore 47 per cent of the cost of the deficiency payments policy, but would have borne only 32 per cent of the cost of EEC policy as it would have applied to Britain in 1969. The poorest quartile, with only 11 per cent of final income, would have borne 13 per cent of the cost of EC policy, compared with only 7 per cent of the cost of deficiency payments. Josling and Hamway then combined their figures for the benefits according to income group of farmers (which were enjoyed disproportionately by high-income farmers) with those of the burdens borne by household income groups. Farm households and other households were ranked by the same income classes. They concluded:

> 'The results are striking. In each case (*deficiency payments, variable levy and EEC policy*) households below an income of £1,850 per year (£36 per week) pay more for farm support as taxpayers and consumers than the farmers in that income group receive. The reverse is true of households with incomes above £1,850 per year. In other words, in all three programmes there is an implied transfer from families below £36 per week to those above that line. In the case of the deficiency payment scheme the size of this net transfer is £67m . . . At the present level of Common Market

THE CASE AGAINST SUPPORT 77

prices the net transfer from those earning less than £36 per week is a formidable £504m of which there is an estimated payment . . . into the EEC Fund of £250m. Thus high-income families receive a transfer of £254m when compared with the absence of support.'[1]

The CAP would therefore imply a very large transfer of income from poor urban households to rich farm households. This 'Robin Hood in reverse' element of the CAP was confirmed and enlarged upon in another recent paper from the Institute of Fiscal Studies,[2] in which the differences between the 'world market' and EC selling prices of various agricultural commodities were treated as a 'tax' and allocated into the intermediate sectors which use those commodities; for example, the support costs of cereals were divided among grain, alcohol, animal feed, and so on. The tax on intermediate inputs was then converted into a tax on final outputs such as bread, milk and milk products, soft drinks, and hotels and catering. The tax rate on final outputs ranged from 0·1 per cent on alcoholic drink, 19·6 per cent on animal feed, to 34·5 per cent on milk and milk products. The total tax was estimated at £2,137 million (1978).

The taxes on final food commodities were then applied to the expenditures of household income groups given in the Family Expenditure Survey to arrive at an implied rate of tax which the CAP imposed on the gross income of each group. Again, the regressive effect came out clearly. The average rate was a surprisingly high 4·5 per cent which ranged from only 2·9 per cent on households with an income of over £10,000 a year to 6·0 per cent on households in the £1,000 to £2,000 bracket. Pensioner households suffered a tax in excess of 5 per cent. The regressive effect is exacerbated because the CAP supports most heavily those commodities (cereals and milk) which go into the products which the poor buy most. When added to taxes as a whole, the effect of the CAP tax 'significantly reduces the progressivity of the overall tax system'[3] in Britain.

Moreover, the consumer of food suffers an additional penalty under the CAP compared with the old British policy. With all its faults, the old system did permit the housewife a wide range of choice of

---

[1] *Op. cit.*, p. 80.
[2] A. W. Dilnot and C. N. Morris, 'The Distributional Effects of the Common Agricultural Policy', *Fiscal Studies*, IFS, London, July 1982.
[3] *Ibid.*

foodstuffs at world prices from all over the globe. Until the early 1960s, nearly every commodity could enter the UK without hindrance from tariffs or quotas, and even until 1973 there were few onerous restrictions on imports. New Zealand and Australian lamb and butter, Argentinian beef, and Canadian wheat were all part of the diet. The CAP, with its variable levies, deliberately gives preference to internal EC production with consequent penalties for imports, the consumption of which has fallen substantially. Ignoring for the moment the adverse effects on the countries penalised and the repercussions for us, the food consumer (which means everyone) is also the poorer through this restriction of choice.

The Benefits to Farmers

As mentioned above, Morris and Buckwell give estimates of 'producer gain' as well as of losses for the consumer and taxpayer. Producer gain is also estimated in the later IFS paper by Dilnot and Morris (1982),[1] who define it as equal to domestic production multiplied by the difference between domestic and world prices (less a deduction for the higher production costs resulting from the higher price of animal feeds) plus direct subsidies. Bowers and Cheshire[2] use a similar definition for what they call 'the total value of producer subsidy equivalent' for 1977, 1978, and 1979. Their figures are considerably higher than those of the other writers – possibly because they choose different world market prices. Their estimate for 1979 (the lower of the two they give) of £2,863 million is double that of Dilnot and Morris, whose figures for producer gain are related to the total cost of support and to farming income (Table XIII).

Two trends can be seen in Table XIII. The first has arisen since Britain joined the EC. Under the old British policy, the whole of the cost of support (apart from administration) took the form of direct transfers to British farmers from British taxpayers and consumers. By 1978, however, as a member of the EC, the total cost of agricultural support to the UK was more than double the gain to UK farmers. Indeed, British consumers and taxpayers paid out over £2,000 million more for agricultural policy than British farmers received from it.

---

[1] *Op. cit.*
[2] Bowers and Cheshire, *op. cit.*, p. 100.

TABLE XIII
TOTAL SUPPORT COSTS, PRODUCER GAIN AND
FARMING INCOME, 1956 TO 1980
(at 1980 prices)

Year	Total Support Costs £m	Producer Gain £m	Farming Income £m
1956	1,867	1,867	1,708
'1966'	1,664	1,664	1,854
1972	n.a.	632	2,097
1976	n.a.	883	2,168
1978	3,598	1,437	1,680
1979	n.a.	1,475	1,344
1980	3,555	1,570	1,018

Sources: Total support costs from Table XI.

Producer Gain: For 1956 and '1966' producer gain has been taken to equal total support cost under the former British policy. The figures from 1972 onwards are from A. W. Dilnot and C. N. Morris, 'The Distributional Effects of the Common Agricultural Policy', *Fiscal Studies*, IFS, London, July 1982. Producer gain is defined in IFS Working Paper 28 (p. 1) as the gain 'to farmers in the form of higher prices received for produce and direct subsidies'.

Farming Income: For 1956 and '1966', the figures refer to NFI and are taken from Howarth, *op. cit.*, p. 36, adjusted to their 1980 value. Figures for 1972 onwards are from Table VIIB.

This shortfall arose partly through the higher prices of the food Britain imported from its fellow EC members and partly through the complex budgetary mechanisms of the Community and the CAP. The £2,000 million went into the pockets of European farmers, to storage agencies, to subsidise surplus disposals, or was otherwise swallowed up by the CAP. From a national standpoint, more than half the cost of agricultural policy is now totally 'wasted' outside the UK – a major factor in this country's long-running battle on its 'budget contribution' to the EC.

The second trend is that, over a long period, 'producer gain' has tended to be roughly equivalent to or in excess of farming income. On his figures for the early 1950s for total support, Professor Nash concluded[1] that 'we reach a total sum which is nearly the same order of magnitude as the aggregate net income of farming in the United

---

[1] E. F. Nash, *op. cit.*, p. 97.

Kingdom . . .' Gavin McCrone's figure for 1956 just exceeded farmers' net income; whilst my own figure for '1966' was a little below NFI. In the early- and mid-1970s, total support costs (for which we do not have specific figures) obviously declined considerably whilst world prices were high; and, briefly, farming income was then considerably in excess (three times) of producer gain. By 1978, however, producer gain and farming income were once more close. The figures for 1979 and 1980 given by Morris, Dilnot and Morris, and Buckwell all show producer gain was by then considerably in excess of farming income. On the basis of Buckwell's figures for 1980, producer gain was £1,904 million (£1,540 million from the CAP and £364 million from national support) and farming income was £1,018 million. Dilnot and Morris's estimate for the same year, given in Table XIII, puts producer gain at £1,570 million.

It is clear that nothing like the whole of the producer gain from support is translated into a net addition to farmers' incomes. Even if producer gain was equivalent to net income, it would not necessarily follow that the abolition of all support would lead to the disappearance of net income, and with it of the whole farming industry – as the defenders of support sometimes seem to imply. That might happen if farmers continued to produce exactly the same output, using exactly the same resources at current prices and exactly the same methods as now. But, of course, none of these would remain the same; after some time for adjustment, they would all change. At world market prices, farmers would produce products in different combinations and volumes, using different amounts of resources the prices of which would also be different. The structure of the industry would change considerably. Farmers did make money in a freely competitive environment prior to the advent of support, and they would make it again after the abolition of support. It follows that, if there would be an agricultural net income without support, and if producer gain were to equal farming income, some of the support would still be 'disappearing' out of farmers' pockets. And when producer gain exceeds farming income, a lot of the support is being dissipated rather than reaching the intended beneficiaries.

Dissipation of support into land values
Where does that part of support which adds to farmers' revenue but not to their net income go? And where has it been going for the

past 30 years or so? There are a number of possibilities: to hired farm workers; to the suppliers of other farm inputs (machinery, fertilisers, chemicals, fuel); or to increase the value of farm land, and thereby the rents of landowners and interest charges paid to lenders. Economic theory teaches that, normally, a subsidy will ultimately be allocated to factors of production according to their elasticities of supply. The gain will be the biggest to those factors with the lowest supply elasticities. Of the possibilities mentioned, land has the lowest elasticity of supply and land values are the obvious candidate to have benefited most from support. They do appear to have done so.

Table XIV shows the movement in average land prices in England and Wales from 1943 to 1983. Land values have increased 40-fold since 1943 and 26-fold since 1953 (when peace-time support really began), during which periods prices generally have risen only 12-fold and 8-fold respectively. The average value of farm land just before World War II was probably about £25 per acre. It is now over £2,000 per acre, or 80 times its pre-war value, whilst prices generally have gone up 16 times. The price of land has increased five times more than retail prices. It would be silly to pretend that the support system has been the only cause of the vast increase in farm land values; there have been several others, including the demand for land as a store of wealth to take advantage of its special treatment under both Estate Duty and Capital Transfer Tax. Ultimately, however, the value of land is closely related to the value of its marginal product and since the price support system has substantially raised the latter, so it has raised land values.

The consequences for farmers have varied according to their circumstances. Tenant farmers have generally gained least from the system, although not during the *whole* post-war period. From 1947 to 1958, rents were held down artificially by legislation and landlords were frequently receiving rents which, after the costs of repairs, provided a minimal or negative return on their capital. Nor were landlords able to benefit fully from appreciating capital values because tenanted land fetched only half or less of the price with vacant possession. They were often forced to sell at low prices to sitting tenants who could immediately enjoy a hundred per cent appreciation of their investment. After the Agricultural Holdings Act of 1958, rents were determined according to a new 'open-market' criterion which led to major increases in later years. This criterion itself is now subject

TABLE XIV
AVERAGE LAND PRICES IN ENGLAND AND WALES,
1943 TO 1983

Year	£/acre	Index (1953=100)
1943	54	65
1947	82	99
1950	95	114
1952	95	114
1953	83	100
1954	83	100
1955	90	108
1956	87	105
1957	84	101
1958	94	113
1959	110	133
1960	132	159
1961	132	159
1962	139	167
1963	168	202
1964	224	270
1965	238	287
1966	246	296
1967	263	317
1968	287	346
1969	302	364
1970	273	329
1971	335	404
1972	555	669
1973	800	964
1974	682	822
1975	584	704
1976	796	959
1977	1,013	1,220
1978	1,365	1,645
1979	1,831	2,206
1980	1,907	2,298
1981	1,831	2,206
1982	1,901	2,290
1983	2,197	2,647

Sources: 1943–1971: University of Oxford, Institute of Agricultural Economics. As quoted in *Farmers Weekly*, 26 January 1973, p. 79.
1972 onwards: *The Farmland Market*, Estates Gazette/Farmers Weekly (various).

TABLE XV

FARM RENTS AS A PERCENTAGE OF
NET FARM INCOME, 1952 TO 1981

Year	%
1952	19·8
1962	27·6
1970	69·8
1975	31·6
1978	42·4
1980	60·1
1981	63·8

Source: M. Murphy, *The Farmers' Club Journal*, No. 67, February/March 1984, p. 45.

Note: Murphy's figures overstate the absolute (but not the relative) percentages in that a deduction for a notional rent has already been made from gross income before arriving at net farm income.

to qualification in a new Act.[1] Having declined after 1958, the premium for vacant possession returned following another Act of 1976 which granted certain rights of succession to tenants (and the provisions of which are now to be partially repealed). The benefits to landlords of rising land values have thus been determined rather capriciously by tenancy legislation. Since 1958, however, tenants have had to pay an increasing proportion of their income in rents, as Table XV shows.

500-acre millionaires

It is the owner-occupier who held land just before or just after World War II who has benefited substantially from the support system. In the 1950s, the farmer with 10,000 acres was a millionaire; by the late 1970s, he required only 1,000 acres; in the early 1980s, 500 acres admits him to that exclusive club. Those who held modest parcels of land in the 1940s and 1950s have been able to borrow comfortably to pay high prices for additional land and have contributed to the bidding-up of land values. Operating on the principle of 'averaging', they might in the early 1970s, for example, have added another 200

---

[1] The Agricultural Holdings Act, 1984, which introduces a new criterion into rent arbitration concerning the 'productive capacity' of the land.

acres at £500 per acre to their existing 200 acres bought at £100, thus creating a farm which, in their own eyes, was worth a mere £300 per acre on average. They seldom stopped to consider what might be the opportunity cost of their £200,000 of capital at that time, or £800,000 at the present time!

The rôle of support in determining land values is evidenced by the fact that landowners benefited from a real capital gain of around 50 per cent in the first 10 years of EC membership between 1973 and 1983 when an even more generous policy operated. That policy also attracted into the farmland market the new class of institutional buyers whose holdings have multiplied by three over the period (although their total holdings are still not large), and whose presence is much resented by many traditional farmers. That infertile land in a 'Less Favoured Area' (discussed in Section VI) which qualifies for special subsidies fetches more than fertile land a few miles away outside the LFA is also illustrative. The present uncertainty in the land market as a result of the introduction of milk quotas and fears of impending quotas or price reductions for cereals further substantiates the link between policy and land values. *The Farmers Weekly* reported on 2 August 1984 that estate agents believed Grade 3 land, which might have sold at £1,800 an acre in 1983, was now unlikely to fetch £1,500 an acre, and that tenants were resisting rent increases to the extent of even asking for reductions on farms subjected to milk quotas.

The diversion of support into land values is a major strand of Richard Body's argument in *The Triumph and the Shame* and one which few, even those who disagree with many of his other points, would dispute. It is a phenomenon which is very similar to the effect on house prices of the relief against income tax granted to house buyers for mortgage interest payments. In both instances government policies have created privileged groups of existing owner-occupiers who have benefited from a huge increase in the value of their property through no effort of their own. In both instances new entrants to the market have suffered from a substantially increased burden of capital costs and interest payments, which in farming has made it virtually impossible for new private entrants to attain the first rung of the farming ladder unless they already possess large sums of capital or other sources of income. For all practical purposes, the ladder is available only to those who are already half way up it. As Christopher

Johnson remarked to the seminar on *The Triumph and the Shame*,[1] borrowing on agricultural land has remained at a remarkably low and constant proportion of asset value over the last decade, suggesting that there has been no rush of new borrowers.

Although land values have pre-empted the major part of support, it has also been diverted into the other factors of production, as is shown in Table XVI which attempts to bring together the pieces in the agricultural jigsaw. Since 1970 farming income has not kept pace with farm product prices, which in turn have not kept pace with the Retail Price Index. Fuel prices have risen fastest of all, as in every industry. They are followed in descending order by land values, farm labour costs, and farm machinery costs, all of which have exceeded the RPI. Despite remaining below average industrial earnings, the earnings of farm workers have improved over the years in relation to both factory workers and farmers. Indeed, if the official statistics are to be believed, the average farm worker now earns considerably more than several categories of farmers. According to *The Economist Diary*,[2] by 1978 the purchasing power of the farm worker's wage was 226 per cent of its 1938 value whilst that of the factory worker was 181 per cent. Most professional people were actually worse off than they had been before the war.

Over-capitalisation

It is also common ground that farming has become over-capitalised. Capital grants have encouraged farmers to purchase buildings and fixed equipment on a lavish scale; and who can blame them for taking advantage of a system which at times reduced the effective cost to them of a new building to only 20 per cent or less of its full cost? Farmers have also been enabled to purchase new machinery in excess of their requirements. There seems to have been a lot of the 'keeping up with the Joneses' motivation behind the purchasing of farm machinery, which has permitted its prices to rise surprisingly rapidly (Table XVI). Again, fears of cuts in support for cereal farmers are currently causing an upheaval in the machinery market and manufacturers were desperately offering generous discounts and special offers during 1984.

---

[1] CEAS/CAS, *op. cit.*, p. 74.
[2] *The Economist Diary 1979*, p. 20.

TABLE XVI

COMPARATIVE INCOME AND PRICE INDICES AFFECTING BRITISH AGRICULTURE (1970=100)

Year	(1) Farm Product Prices	(2) Farming Income	(3) Land Values	(4) Fuel	(5) Farm Machinery	(6) Farm Labour	(7) Retail Price Index
1964/65 to 1965/66	90	75	76	78	81	65	78
1970	100	100	100	100	100	100	100
1971	106	113	111	103	108	109	109
1972	112	120	183	107	118	122	117
1973	147	168	264	122	131	150	132
1974	166	142	225	179	159	195	154
1975	206	177	193	216	200	245	191
1976	265	228	263	266	235	293	222
1977	275	224	334	320	291	318	258
1978	294	221	450	328	339	365	279
1979	313	201	604	393	377	425	317
1980	331	179	629	515	435	516	374
1981	366	232	604	623	470	570	418
1982 (provisional)	392	318	627	702	507	623	452

Sources: *Annual Review of Agriculture 1983*, Cmnd. 8804, HMSO, 1983; and previous Annual Review White Papers.
Land Values: From Table XIV, the 1970 base=the 1969-71 average.

Gross fixed capital formation in agriculture (before the deduction of grants) has been rising more rapidly than the sector's contribution to GDP. In 1983, the former amounted to 320 per cent of its 1972-74 figure whilst the latter had risen by 301 per cent.[1] Capital formation in plant, machinery and vehicles rose by 354 per cent over the same period. This investment has been reflected in a remarkable growth in borrowing from the banks, which has exceeded the rate of inflation. Total borrowing from the banks by UK agriculture amounted to £4,699 million in 1983, which is more than seven times the nominal figure for 1972 and represents a real increase of well over 70 per cent.

---

[1] Cmnd. 9137, *op. cit.*, p. 37.

Total advances to agriculture are equivalent to between one-fifth and one-quarter of all advances to manufacturing industry, although the contribution of agriculture to GDP is only one-tenth that of manufacturing. Some people might regard this debt exposure as potentially dangerous. However, the total asset values of agriculture, of which farmland and buildings comprise three-quarters, are so large that, according to Johnson,[1] 'agriculture was more highly geared in 1970 than in 1982, with liabilities falling from 15 per cent of assets then to 11 per cent now'. On the other hand, that low gearing is obviously very sensitive to changes in land values, which in turn are sensitive to changes in support policy. Hence at the moment, when there are fears of changes in the amount of support, propositions from farmers are being considered much more cautiously by the banks than for a long time.

For many, if not most, farmers, assistance in the form of general price supports and input subsidies has not provided the income bonanza which their critics often accuse them of having enjoyed. A large proportion of support, apart from that which leaves Britain under EC policy, does not end up as net income for farmers. It disappears into the pockets of the suppliers of farm inputs, into the profits of the banks (interest on commercial debt, excluding loans for land purchase, is now equivalent to around one-third of farming income), into the returns of landlords from rent, and into the store of wealth of land-owners. As already mentioned, the relative incomes of farmers have not kept pace with other incomes despite a doubling of the real burden of support on consumers and taxpayers since the 1950s and 1960s. Farmers have been worshipping at the altar of a false god who has showered his bounty on them in an extremely haphazard manner.

Paul Cheshire was right in his comment on television that to own (or to rent) agricultural land is to own a 'licence to receive subsidies';[2] but it is not necessarily a licence to receive a high net income. There are undoubtedly some farmers with high apparent incomes of £30,000 a year or more (plus other benefits) who are enjoying a high standard of living. When, however, their income is related to their capital, they are probably earning a return of only 3 to 5 per cent, even in

---

[1] C. Johnson, in CEAS/CAS, p. 74.
[2] In *Against the Grain*, Thames TV, 26 July 1983.

good years. They have made spectacular capital gains if they have owned the land for long enough, and as individuals they may be extremely wealthy. But they are wealthy only so long as (a) the support system continues and (b) very few of them wish to dispose of their land at a particular time. Land values are high because land is largely fixed in supply and there has been a keen demand for the very small proportion of it which comes onto the market each year. What would happen to land values if, say, 25 per cent of the total acreage was put onto the market in a single year? No-one can predict accurately, but most people would anticipate a very substantial fall in its price.

Balloon theory

I have long held a theory about agriculture – I term it the Balloon Theory – which states that, just as it is impossible to inflate one side of a balloon without inflating the other, so it is impossible artificially to inflate the revenue of farmers without inflating their costs. The agricultural balloon may be a bit lopsided in that the revenue has been inflated somewhat more than the costs, but most of the money pumped into one side has ended up in the other. Some of the smaller and marginal-land farmers have been enabled to stay in business by specific direct grants and subsidies when they would otherwise have had to give up. Some of the larger farmers who either own their land or are efficient tenant farmers on good cereal-growing land have done very well. But the mass of farmers, in the middle, have probably gained little out of the support system. Their very strong desire to enter or remain in farming and their acceptance of modest incomes in return for the non-pecuniary attractions of farm life have made farmers willing to pay high rents and high interest payments on land purchases which have largely offset the benefits to them of the support system. So long as such people exist, their actions will tend to offset the *gross* benefits from price, input, or income support, however high they may be. The experience of farming over the period of support confirms the notion that, in Britain, the USA and Western Europe, there is a 'natural level of disparity' of income between agriculture and other sectors (which may be close to parity, as in Britain, or around 50 per cent, as in other countries) which it is impossible to alter except during occasional brief interludes after support has been substantially increased. This notion is akin to that of the 'natural rate of unemployment' which is equally resistant to modification (in anything other than the very short term) by the application of government policy – in this instance, reflation.

Richard Body argues that the 'shame' of it all is the enormously increased burden of support and the apparent growing dependence of net farm income on it. On the basis of the 1980 figure in Table XIII, total support now works out at £17,000 per full-time farmer, nearly £12,000 per farmer if part-timers are also included, and almost £5,500 for every person engaged in agriculture. It is also a shame for the rest of us that support has done farming so little good over the long run. This is not simply a view deduced from personal observations, it is confirmed by econometric studies by Bruce Traill of Manchester University.[1] He has calculated that a 1 per cent increase in support prices in Britain raises net farm income by around 9 per cent after time for adjustment. 'However, most of the benefits . . . become capitalised into land values, which are also seen to rise by about 10 per cent (£130 per acre) in the long run'. Investment predictably rises by a cumulative total of £44 million (representing an increase of about 0·4 per cent in the capital stock). Surprisingly, however, the employment of hired labour is reduced by around 1 per cent. Thus support can produce an effect precisely opposite to the declared aim of the CAP to retain labour in agriculture.

Economic Costs

Richard Body seeks to demonstrate that, apart from hurting the taxpayer and the consumer, agricultural policy has had severe adverse effects on the economy as a whole through the 'diversion of capital'. It is at this point that his arguments become least plausible; he overeggs the pudding by confusing capital formation with increased land values and by equating loans to agriculture with loans to the industries which serve agriculture. He asserts that the capital invested in land by institutions (which is a minute fraction of their total investment) and the investment by chemical companies in production

---

[1] W. B. Traill, 'Land Values and Rents: The Gains and Losses from Farm Price Support Programmes', *Department of Agricultural Economics Bulletin 175*, University of Manchester, 1980; also Traill, 'The Effect of Price Support Policies on Agricultural Investment, Employment, Farm Incomes and Land Values in the UK', *J. Agric. Econ.*, Vol. xxxiii, No. 3, September 1982.

facilities for fertilisers and farm chemicals have starved the remainder of British industry of investment funds and thereby lowered its performance. That assertion is untenable. Some Labour politicians used to speak of an 'investment strike' by banks and financial institutions which were failing to provide British industry with much-needed capital. This myth, however, was laid to rest by the Wilson Committee[1] which found there was a plentiful supply of funds available for viable projects. It was good investment projects which were in short supply, not the finance for them. The apparent vast capital locked up in land is mainly a 'paper gain' in the hands of many individual landowners and can never be available for re-investment elsewhere, except when released in very small slices.

The relatively poor post-war performance of the British economy has not been caused by farm policy. Other countries such as France, West Germany, Sweden and Japan have subsidised their agriculture to an even larger extent, yet their economic performance has outstripped Britain's. This is *not* to deny, however, that farm support has caused a considerable misallocation of resources and a loss of welfare. The money diverted to agriculture from consumers and taxpayers through higher food prices and taxes could have been left to 'fructify in the pockets of the people' – to be spent by them as they chose on other goods and services or investments from which they would have gained more satisfaction. The loss they have suffered is not once-for-all; it is a cumulative one over a period of 30 years and is difficult to measure. Furthermore, it is impossible to measure the loss (which could be substantial) caused by the diversion of entrepreneurial talent into agriculture.

In his review of Body's book, George Peters[2] maintains that the over-capitalisation of farming has caused a minimal loss to GNP and economic growth, which 'merely serves to emphasise the fact that what may be a large misallocation in relation to a small sector may have trivial overall effects'. Similarly, the agricultural lobby argues that since many other industries are subsidised and protected, it is unfair to single out agriculture as being particularly favoured: 'if other people are getting support, why should not we?' Indeed, there

---

[1] *Final Report of the Committee to Review the Functioning of Financial Institutions* (Wilson Committee), Cmnd. 7937, HMSO, 1980.
[2] G. H. Peters, *op. cit.*

has been a propensity in recent months to compare agriculture with the coal industry, as Michael Murphy[1] did in an address to the Farmers Club. Jock (Lord) Bruce-Gardyne also drew a comparison in the *Sunday Telegraph*, albeit to advance a quite different argument:

'... What is the merit in handing over taxes to pay middle-aged men a pittance to go out in rain and snow to cultivate a Welsh mountainside which could more gainfully be left to grouse or trees, or to keep other middle-aged men burrowing into the earth for coal which the Coal Board and the rest of us would be better off without?'[2]

And although Arthur Scargill and Anthony Wedgwood Benn naturally view matters differently, both asked on various occasions during the 1984 coal strike why Socialist miners should not receive subsidies when Tory farmers do. And, indeed, why should they not?

The ideal is that neither should receive subsidies. Although the distortions caused by support for an individual industry may be 'minimal', the aggregate effect of assistance or protection for *all* industries must be appreciable. Here is an argument which *is* relevant to Britain's economic performance. The rigidities and misallocations which beset the British economy are so numerous as to make it arthritic. The precondition for fundamental improvement is to eliminate the obstacles to a more efficient allocation of economic resources which are erected and buttressed by the supporters of all political parties. It is hazardous for people in glass houses to throw stones. Politicians who wish to reform trade unions and curtail their restrictive practices, and who seek to reduce subsidies to nationalised monopolies and return many of them to private ownership, must at the same time be willing to attack the monopolies, restrictive practices and subsidies enjoyed by their own supporters – be they opticians, solicitors, barristers, accountants, architects or, above all, farmers. It is in this sense that farm support is the biggest hindrance to economic progress.

Further Observations on the CAP

So far, the arguments against supporting farm prices and inputs have been concerned with their general effects on farmers, consumers, taxpayers, and the national economy. The CAP, however, has addi-

---

[1] M. Murphy, *op. cit.*   [2] *Sunday Telegraph*, 3 June 1984.

tional adverse consequences. Since so much has been written about the CAP, it is not necessary to describe it in detail here.

Internal farm prices raised

By having a common policy for agricultural prices, the EC has raised the internal prices of farm products to a higher level than they would otherwise have reached under the national policies of most member countries.[1] This became apparent in the mid-1960s when the common cereal prices were fixed, after long political wrangles, well above the previous average in the then six member countries. Cereals gave the lead for other products. Farmers in the Netherlands, Belgium and France were given an immediate price boost which much more than outweighed the small reduction in Germany, Italy and Luxembourg where farmers received special compensation. A similar boost was given to Denmark, Ireland and Britain on their entry in 1973. Many commentators would agree that, with the possible exception of Germany (which has long maintained a positive MCA and thus kept its farm prices above those in other EC countries), no member country would be willing to bear individually the cost of supporting its farm product prices at current EC levels. The CAP has thus exacerbated the internal costs and problems of price support.

The scope for fraud

A second, perhaps minor but often neglected, consequence of the CAP is that it has considerably increased the scope for fraud in claiming agricultural subsidies. Not only is the European system much more complex than the national policies it replaced; it also involves numerous categories of persons other than farmers who can make dishonest profits out of it. A few of the potential areas for fraud are: sales of cut-price butter to less-favoured groups in the population; deliveries of cut-price foodstuffs to the armed forces; the distillation of wine surpluses; the subsidising of food aid to the Third World; and the supply of food aid to disaster areas. Farmers themselves have found numerous ways of cheating, including false claims for slaughter

---

[1] In 1981 Community butter prices were 53 per cent higher than on the world market, beef prices 52 per cent higher, wheat 38 per cent, barley 35 per cent, sugar 33 per cent, and pork 24 per cent. (*The Economist*, 23 October 1982, pp. 54, 55.)

subsidies for dairy cows and grants for the conversion of dairy herds to beef production. The biggest illegal profits, however, have probably been made out of the green money system by cross-border smuggling. The *Economist*[1] has described how it is done:

'When Britain had a negative MCA, for instance, Irish farmers could load their cereals into trucks, drive into Northern Ireland and collect an import subsidy. They would then drive out again along a road without border controls, pick up new papers and drive through again to collect another MCA. The swing to a British positive MCA simply meant that the smuggling had to be in the opposite direction. Belgian farmers have copied this wheeze by walking their cattle or pigs into Holland at dead of night, having them slaughtered there (to disguise the livestock's origin) and then importing the meat back into Belgium to pick up an 8·5 per cent subsidy (the sum of the Belgian and Dutch MCAs).'

The continental press has reported more spectacular frauds involving traders with barge-loads of butter and grain who have crossed frontiers collecting MCA subsidies, and have then gone back with the same consignments disguised by a light covering of coal only to return with the original produce to collect further MCA subsidies. An EC Commission newsletter on the CAP[2] contains five other examples of ingenious frauds which have been detected. It does not mention the many other possibilities, but they are legion.

No-one knows the full extent or the cost of criminal activity within the labyrinth of the CAP but the Commission is concerned about it. Under the Rome Treaty the Community (that is, the Commission) has no power in this matter; it is the responsibility of national law enforcement agencies to catch and take legal proceedings against offenders. The problem is that individual national authorities tend to take a more relaxed and lethargic attitude to policing offences against the Common Market and to tightening up procedures connected with it than towards, say, income tax evasion in their own country. When the loss is incurred by a general fund financed by everyone it is perceived to be less serious than a loss to the national exchequer, and the incentive to do anything about it is smaller.

The corruption which it inevitably breeds is an argument against

---

[1] *Ibid.*, p. 55.
[2] *The Prevention of Frauds Against the Agricultural Fund*, Newsletter on the CAP, No. 193, EC Commission, 1982.

any bureaucratic system of subsidy and control. It is a further charge in the indictment of the CAP. But more serious charges, perhaps the most serious, relate to its external consequences.

Effects on the Third World

The EC Commission is very sensitive to the charge that the CAP harms the agricultural exports of Third World countries. It points out that the Community remains a net importer (and the world's largest gross importer) of agricultural products, and that almost all its imports from the 66 countries of the African-Caribbean-Pacific (ACP) group enjoy free access to the Common Market under the Lomé Convention. It neglects to mention that most of these imports are in the wide range of tropical products which cannot be produced in Europe and that there is therefore no conflict with EC producers' interests in allowing them free entry. There are, however, certain commodities the Third World exports which do have to compete with CAP régimes. They include sugar, tobacco, cotton, olive oil, fruit, vegetables, beef and wine. Some of these products enjoy preferential access to the Community's market, but most have to compete against CAP export subsidies in trying to find other markets. An American study has shown that even a 50 per cent reduction in tariffs by the OECD countries (of which the EC members are the most serious offenders) could lead to substantial increases in the export revenues of the less-developed countries (LDCs) – as high as 74·9 per cent for beef, 46·1 per cent for refined sugar, and 46·3 per cent for wine, with smaller increases for several other commodities.[1] The Community's sugar policy inflicts the most damaging losses on the potential export revenues of LDCs.

It can be argued, on the other hand, that the LDCs benefit as food *importers* from the support policies of the Community and other OECD countries since those policies reduce the demand for temperate farm products in the developed world and create surpluses for export (frequently subsidised) which depress world prices for temperate food. This is true of both grains and dairy products which are imported in large quantities by LDCs and whose consumers can enjoy higher

---

[1] Figures quoted in D. R. Colman, *op. cit.*, p. 53, and taken from A. Valdes and J. Zeitz, *Agricultural Protection in OECD Countries: its Cost to Less-Developed Countries*, International Food Policy Research Institute, Research Report 21, Washington DC, 1980.

consumption than they would otherwise be able to afford. In 1984, the LDCs will probably have imported four times the amount of cereals they did in 1960 – a rate of growth twice that of their population. Commercial imports will have accounted for over 90 per cent of total imports (at a cost of $21 billion); the remainder will have been food aid.[1] The whole of this increase will have come in the better-off developing countries. Most Latin American and Middle Eastern countries are now heavily dependent on imported grain, which has enabled them to move farmers into new industries. But in black Africa and most of Asia, the people are largely dependent on food produced by their own farmers who constitute the bulk (maybe 70 per cent) of the population. The small proportion of the population which lives in urban areas may benefit from lower import prices. However, the general development of these countries is retarded because the agricultural policies of the developed world undermine local farmers – as do their own governments which frequently exploit them by forcing them to sell to state marketing boards at less than world prices and which keep the cost of imported food artificially low by maintaining over-valued exchange rates. At the present time the activities of their own governments probably constitute the biggest threat to agricultural progress in such countries. More important in the long run, however, as Enoch Powell has rightly remarked, is the '... paradox that we starve those countries precisely by refusing to buy food from them, when food production for export is their only access to wealth and to international purchasing power'.[2]

The persistent refusal of the EC to put into practice its professed belief in 'trade not aid' was demonstrated in October 1984 during the negotiations for the Third Lomé Convention when it had the opportunity to expand its trade with the poor countries. Despite the protests of the other Community countries, those whose farmers would suffer most from increased competition in Mediterranean-type goods

---

[1] There will always be a requirement for food aid to relieve disasters (as in Ethiopia and other African countries currently), but the Green Revolution is spreading rapidly in China, India, Latin America, and South-East Asia. Most of black Africa (Kenya is an exception) has a long way to go in catching up. Political mismanagement rather than lack of the appropriate technology is the prime problem there. But the total requirement for food aid is likely to decline and therefore to remove the basis of the simplistic argument that it would be immoral for the developed countries to curb their food production when so much of the world is hungry or suffering from malnutrition.

[2] J. Enoch Powell, *The Spectator*, 9 June 1984, p. 27.

refused to expose their markets to more ACP exports. Minor concessions were made for beef, rice, and rum; and cash aid was increased in line with inflation. But ultimately, as *The Economist* put it: 'It was all the ACP countries could do to stop the Community back-sliding from its commitments under the current Lomé convention, number two, which runs out in February'.[1]

Trade wars

In 1979, for the first time, the Community became a net exporter of temperate foods, and by far the second biggest food exporter in the world after the United States. In 1981, the EC exported more cereals than it imported and now has surpluses of wheat, barley, sugar, pork, beef, wine and, above all, dairy products.[2] Yet its internal prices remain as much as 50 per cent higher than world market prices and its exports have to be heavily subsidised. This dumping on world markets absorbs half the total expenditure (estimated at £9·6 billion in 1983) of the EC's agricultural fund (FEOGA).

The CAP's objective of self-sufficiency, promoted by high internal prices and variable import levies, has long been a bone of contention with smaller countries which depended heavily on their agricultural exports to Britain and other Community countries. Traditional suppliers in this category include New Zealand (lamb, butter and other dairy products, beef), Australia (cereals, meat and sugar), Canada (wheat) and Argentina (beef). Today, however, not only are they still suffering from the loss of the European market; they are also facing subsidised competition from EC agricultural products in their other export markets. This is depressing the prices they receive and is financed by export subsidies which they could not possibly afford to match.

The United States can afford to match them and to top them. Until very recently, America had been a tolerant and apparently slumbering

---

[1] *The Economist*, 13 October 1984, p. 66.
[2] The cost of supporting the wine market is expected to have exceeded £600 million in 1984. Up to one-third of the wine produced each year has to be converted at a loss into industrial alcohol. The butter mountain has risen from 900,000 to 1,200,000 tonnes, which could cost an additional £600 million in export subsidies if a market for it could be found. And if the milk quotas are successful in reducing dairy surpluses, they are likely to be transformed into beef surpluses as part of the dairy herd is slaughtered and producers move into beef production. Beef stocks already exceed 300,000 tonnes.

giant. It had complained, but done little else,[1] about the gradual loss of its grain markets in Europe as the Community gradually became self-sufficient, or about its losses in other markets (especially in the Third World) as European exports expanded. In 1983, however, the prospect of having nowhere to store or export a large part of its wheat surplus stung the Americans into action. Total federal farm support, which had been $15 billion in 1982, was increased to $22 billion. The additional expenditure went into more programmes for reducing crop acreages – including the PIK (payment-in-kind) programme under which farmers were compensated for ploughing up crops with gifts of grain out of public stores (82 million acres were 'set-aside' in 1983) – and into the large-scale promotion of export sales. To teach the Europeans a lesson, the Americans 'stole' a long-standing French contract to supply flour to Egypt by heavily undercutting the price. President Reagan told the annual conference of America's main farm union, the Farm Bureau, that if the Europeans wanted a 'food war' he had the will and the resources to fight one and win it.

Fortunately, although the US and EC have fought an inconclusive battle of complaint and counter-complaint in the GATT since 1981, a general food war has not yet begun. Indeed, it has been averted by the poor Canadian harvest in 1983 and by increased Russian purchases of American grain since June 1984 which had already reached 14 million tonnes by end-1984 and may rise to 20 million tonnes in 1985.[2] But the threat of such a war still hangs over us, with the danger that it could spill over into retaliatory action in industrial trade and thus precipitate a general outbreak of protectionism. This possibility could be described as a ride for all of us on a 'hay wain into hell'.[3]

The USA and the EC have for several years engaged in a slanging-match about whose farm policy is the bigger offender in damaging the prospects for free world trade. Both parties spend vast sums of taxpayers' money on farming. In 1984, the Community will have spent $16 billion under the CAP and its member countries will have

---

[1] There had been minor skirmishes in 1963 over the decline in American chicken exports to the EEC which led to reprisals against certain European exports to the USA (the 'chicken war'). And, in 1976–79, a 'turkey war' of similar origin was fought.
[2] *The Economist*, 15 September 1984, p. 45.
[3] This graphic phrase was used as the title of the leading article in *The Economist*, 14 April 1984.

spent a similar sum on national policies – making a total of, say, $32 billion. To that figure must be added the costs to consumers of higher European food prices which, using Morris's estimate,[1] may amount to another $18 billion. The grand total of the cost of support will thus have been of the order of $50 billion. American federal expenditure of $22 billion also seriously understates the total farm support, which the Chemical Bank has calculated at $60 billion. On these figures, American farming is even more heavily supported than European – strikingly so when the cost is spread among the much smaller American farm population. For each member of the agricultural working population (2·2 million in the USA and 8·7 million in the EC in 1980),[2] the Americans spend $27,000 on support and the Europeans $5,700.

However, the way in which the European farmer is supported is much more protectionist. The main differences in the method of support are, first, that the American farmer is paid close to the world market price for the products he sells (and also in the compensation he receives for the products he refrains from producing). Secondly, much American policy is aimed (and has been since 1933) at reducing output through: acreage and marketing quotas for the main crop products; loans to farmers on the security of their produce, subject normally to participation in a 'set-aside' programme; 'rental' payments for the long-term withdrawal of land from production and using it for 'conservation'; and special emergency schemes to reduce output drastically such as the feedgrains programme of the Kennedy Administration and the 1983 PIK programme. In contrast, European policy has encouraged increased output by maintaining high support prices which have for the most part been 'open-ended', that is, paid out whatever the quantity produced.[3] The only restrictions have been quotas on sugar beet and, very recently, milk.

There can be little doubt that, in grains at any rate, the American farmer is highly efficient and could favourably compete on equal

---

[1] C. N. Morris (1980), *op. cit.*, p. 28.
[2] *Agriculture in the United States and in the European Community: a Comparison*, Newsletter on the CAP, No. 200, EC Commission, 1984, p. 3.
[3] Britain is now a contributor to the EC grain surplus. Whereas British farmers sowed 2 million acres of wheat in 1969, in 1984 they sowed 4·9 million (an increase of 16 per cent on 1983). The Intervention Board for Agricultural Produce, which took in 131,000 tonnes of wheat in 1983, was planning to store over 1 million tonnes in 1984. Total EC wheat output has risen from 40 million tonnes in 1970 to 60 million (estimated) in 1984.

terms with Europe or anywhere else. He has the comparative advantages over most of his European competitors of geography, farm size and technology. He is the more justified in his sense of grievance (which is not to say that American farm policy has been sensible from the standpoint of the rest of the population). The US policies of encouraging people to stay on the land by paying them generously not to produce have equally perpetuated a 'farm problem'; Americans, too, have a butter mountain, stored in vast caves under Kansas City, because they have been lavish towards their dairy farmers.

The agro-political dimension

The food surpluses of Europe and America have become a source of political embarrassment and a political bargaining counter. The United States has long been a major supplier of grain to Russia, raising the question whether it is sensible to provide the staple food of a potential enemy and thereby enable him to continue to neglect his agriculture and divert resources to armaments. It has also raised the possibility of using 'food power' as a political weapon with which to punish Russia for its most offensive acts. President Carter attempted a grain embargo on Russia in January 1980, following the invasion of Afghanistan, at a time when the Soviet Union had made plans to import a record 35 million tonnes of grain during that year. The embargo proved to be ineffective because other suppliers (such as Canada, Australia, France, Argentina and South Africa), acting treacherously in American eyes, made up the 17 million tonnes which Carter had hoped to withhold from Russia.[1] It also immensely upset the American farm lobby and proved a factor in Carter's defeat in the subsequent Presidential election.

While the Americans have provided Russia's bread, the Europeans have increasingly supplied both its bread and butter, despite bitter protests from many quarters within the EC. The Community, too, tried to punish Russia over Afghanistan with a butter embargo in 1981 and 1982; and it, too, failed because traders found ways round it. The British Intervention Board for Agricultural Produce, for example, authorised a £4 million export subsidy on butter which,

---

[1] R. L. Paarlberg, 'Lessons of the Grain Embargo', *Foreign Affairs*, Vol. 59, No. 1, 1980, pp. 144-162.

though believed to be destined for Cuba, probably went to Russia. That matter is part of an investigation by the EC Commission into illegal butter shipments during the embargo.[1] Since then, sales to Russia have resumed; and not only of butter, but also of wine, beef, sugar, and grains.

The scope for, and wisdom of, using food sales as a political weapon are now very much in doubt. However, it is strange, to say the least, that America continues to provide so much of the Communist world's food, even at commercial prices. Unambiguously crazy is the European subsidisation of food sales to the Eastern bloc to the extent that Eastern bloc governments, if not consumers, can buy those products at half or less than half their price to the Western European housewife.

The destruction of the countryside

It is impossible to discuss the arguments against farm support these days without mentioning conservation, since the conservationists have come to the forefront in recent years as the attack on farm policy in general and the CAP in particular has mounted. Because, however, this aspect has been well covered in the recent book by Bowers and Cheshire,[2] a few observations only will suffice here.

First, many conservationists are not attacking agricultural policy as such, but modern farming. They object to the removal of hedgerows and ditches; the use of artificial fertilisers, herbicides, and chemicals generally; drainage and the reclamation of scrub and woodland; and the excessive use of farm machinery which consumes 'scarce' fossil fuels. The most fundamentalist of them have the totally unrealistic objective of re-creating a landscape as it may have been 100 or so years ago, with 'organic' farming methods (no artificial fertilisers or chemicals) and the replacement of diesel power by reinstating horse power. How they would persuade or compel farmers to do these things, and what the consequences for our food supplies would be if they succeeded, are another matter.

Secondly, it is true that farm policy, with its high output prices and subsidised inputs of fertilisers, machinery and fuel, has encouraged the expansion and intensification of crop production – frequently on

---

[1] Report in the *Sunday Telegraph*, 15 July 1984.
[2] Bowers & Cheshire (1983), *op. cit.*

land which has been drained and reclaimed with generous grants. It can be assumed that, with lower output prices, fewer fertilisers and chemicals would be applied to land used for cropping; and in the absence also of special subsidies and 'reclamation' grants much marginal land would be withdrawn from cultivation because it would no longer be economic. To the extent that the CAP has promoted higher prices than otherwise and has been especially generous to cereals, it can be indicted for aggravating 'undesirable' changes in the countryside which were already being encouraged by the old British policy. Here, the market economist and the conservationist are on common ground in advocating a reduction in agricultural support.[1]

Thirdly, if farmers are to be paid compensation out of the public purse (which itself is questionable) for keeping their land in the state the conservationists wish, that compensation will be inflated if the value of the production foregone is valued at support prices – as it has to be under the present Wildlife and Countryside Act (1981). Support policy thus makes a conservation policy that much more expensive.

Fourthly, it is relatively easy to value the *costs* of conservation in terms of the agricultural production foregone (the real cost to society being found by valuing it at world market prices), but how are the *benefits* to be valued? What is the value to the individual of looking at a more pleasing countryside and how many individuals are going to see it and be pleased by it? What is the benefit of preserving a species of butterfly, or bird, or grass? To whom does that benefit accrue? The sophistry of welfare economists may offer theoretical answers to some of these questions, but how are they to be applied practically? And if the benefits of preserving endangered marshes and wildlife species are appreciated only by a small number of people who care about such things – the rest of the population being indifferent – why should taxpayers generally (that is, everyone) have to meet the cost?

Fifthly, if the aim is to preserve a relatively small number of Sites of Special Scientific Interest (SSSI), why should the owners of the land be paid compensation at all for not changing its state? Planning

---

[1] A well-expressed conservationist view of the effects of policy in expanding the corn acreage, increasing the use of chemicals, and intensifying livestock production with compound feeds (rather than grass) is given in G. Harvey, 'The Great British Farm Waste Scandal', *New Scientist*, 23 September 1982, pp. 820–824.

restrictions on the erection and alteration of buildings, and preservation orders on trees and historic buildings, are already accepted as public goods. In both instances, the owners of land and buildings are forbidden to do what they wish with their own property. Why should it be different for the owners of certain marshes, fields and hedgerows?

Sixthly, the ecology bandwagon is already rolling rapidly. On to it have jumped an assortment of groups ranging from Greenpeace and Friends of the Earth, the Ramblers Association and the Royal Society for Nature Conservation, to the House of Lords and the Bow Group. Now the National Farmers Union and the Country Landowners' Association are trying to climb on board. They see grants for conservation either as an additional source of funds from the public purse or as a full replacement for existing funds which they fear might be reduced. There is a danger that this bandwagon will become another gravy-train for farmers, with all the attendant consequences of the other types of support we have discussed.

VI. The Politics of Agriculture

The previous discussion has examined agricultural policy from the standpoint of an agricultural economist concerned with such economic criteria as the efficiency of production in meeting consumer preferences. But, as was said earlier, agricultural policy is influenced by political calculations from which agricultural economists have tended to shy away. Indeed, if we want to find the real explanation for the continuation of economically costly and inefficient policies, we need look no further than the politics of agriculture, which provides an excellent illustration of two topics discussed in previous *Hobart Paperbacks* – the operation and motivation of bureaucracy,[1] and the vote motive.[2] In the interests of brevity, this Section focuses on the politics of British agriculture, but the principles apply equally to the USA and our EC partners. There are many aspects of the politics of agriculture, some of which need further research. My argument is confined to four aspects: the general climate of opinion in which agriculture functions; the agricultural interest groups; the agricultural vote; and agriculture in Parliament.

The General Climate of Opinion

The effectiveness of any group seeking political favours to achieve its objectives will depend on whether the climate of opinion in which it operates is favourable, unfavourable or neutral to that group. This in turn will depend upon the existence and effectiveness of any countervailing groups. If no offsetting influences exist, it is likely that the capacity to harm the rest of the population will be biggest. If there are opposing groups of equal strength, the effects will tend to cancel out. Examples of countervailing strength might be the road *versus* the rail lobby, the gas *versus* the electricity lobby, and the navy *versus*

---

[1] W. A. Niskanen, *Bureaucracy: Servant or Master?*, Hobart Paperback No. 5, IEA, London, 1973.
[2] G. Tullock, *The Vote Motive*, Hobart Paperback No. 9, IEA, London, 1976.

the airforce lobby. There are also individual employers' organisations *versus* individual trade unions, and the TUC *versus* the CBI.

However, the agricultural interest groups had a very long, largely unopposed run in Britain and elsewhere which enhanced their inherent strength. They were unopposed for a number of reasons. First, until Britain joined the EC, its agricultural policy was invisible to the mass of the people; it was mainly financed out of general taxation and it did not raise the price of food to consumers or accumulate large, photogenic stockpiles of surplus products. It was therefore of little interest to the mass media. The CAP has since changed all that. Prior to 1973, however, *The Economist* was the only significant newspaper to have consistently condemned the support system from the late 1940s.

Secondly, academic opposition was also slight, which is not to belittle the courageous efforts of a few. The problem was, and to a lesser extent remains, that general economists tended not to interest themselves in agricultural matters which they left to the agricultural economics profession. That profession has consisted mainly of members of departments of agricultural economics which have had very close links with the Ministry of Agriculture in carrying out the Farm Management Survey. The staff of the survey were originally employed directly by the Ministry but have since worked for it indirectly under contract. Many of the people who called themselves agricultural economists have therefore been farm economists working in the field of farm management and closely allied subjects – and have been financed by government. They had neither the training nor the inclination nor the motivation to question the broader issues of policy. And it was not in the interest of other members of those agricultural economics departments to antagonise either the bureaucrats and politicians of the Ministry, for fear of depriving their departments of funds and prestige, or the farmers, for fear of losing their co-operation in the survey.

The exception to the rule was Aberystwyth where the late Professor Eric Nash, himself formerly a civil servant in the Ministry, became one of the foremost academic critics of official policy from the late 1940s to his death in 1962. He attracted like-minded colleagues and students to his department and, in the early 1960s, they formed a distinctive and dissenting group which included Gavin McCrone, Gwyn James and Graham Hallett. On many occasions they came under fire from the local National Farmers Union for expressing views inimical to it.[1] The Union obviously considered that Nash and colleagues were employees of the Ministry who should toe the official line. Indeed, the Union has had considerable success in 'infiltrating' the universities by sponsoring a Centre for European Agricultural Studies at Wye College, University of London, and the Chair of Agricultural Marketing at Aberystwyth. The then chief economist and Deputy Director General of the NFU, Asher Winegarten, was appointed visiting Professor at Wye College in 1975. There has also been a considerable interchange of personnel between university departments of agricultural economics and the economics and statistics sections of the Ministry of Agriculture, further illustrating the inbred nature of the profession.

The other centre from which consistent academic opposition to policy has come is the Oxford Institute of Agricultural Economics (which is completely independent of the Ministry) and especially in the 1960s under the Directorship of Dr Colin Clark. By and large, however, the agricultural economics profession has steered clear of direct opposition to agricultural policy,[2] preferring to remain hidden behind the covers of the *Journal of Agricultural Economics* which, whatever its merits, cannot be said to be widely read. Much of the material for an indictment of policy is in it, but members of the profession have been reticent in taking a firm and unequivocal view.

Only in the past few years, as the excesses of the CAP have become more obvious, has agricultural policy attracted media coverage and attention from outside commentators, with the result that the climate of opinion is more sceptical than ever before and the farm lobby is for the first time having to defend itself to the outside world. Its inexperience of presenting its case in a questioning and occasionally hostile atmosphere has become painfully obvious. Both consumer groups, such as the Consumers' Association, and environmentalist groups have developed strongly in the 1970s and provide an opposing view to the farm lobby. The farmers are fortunate, however, in being

---

[1] G. R. Allen, 'The National Farmers Union as a Pressure Group', *Contemporary Review*, May and June 1959, tells of the problem Professor Nash encountered from the Montgomeryshire NFU when he spoke out against the proposed Egg Marketing Board in 1956.

[2] There are several academic opponents of the CAP but their arguments are based mainly on consumer, budgetary and environmental costs and they tend implicitly to assume that the UK would go back to deficiency payments if the CAP broke up.

only one of many targets for the attacks of those groups, whose fire is therefore less concentrated.

The Agricultural Interest Groups

Despite the growth of opposition, the farm lobby remains one of the most solid and deeply entrenched pressure groups in this country – and probably more so in the USA, France and West Germany where farmers are proportionately more numerous. Traditionally, there have been what *The Economist* has called 'three wings' to the farm lobby – the farmers' organisations, the workers' organisations, and the MAFF itself. More recently, however, a new wing has grown up which might be termed the ancillary industry lobby.

Farmworkers

To start with the weakest, the workers hold an historic place in trade union mythology because of the Tolpuddle Martyrs (1834), but their National Union of Agricultural and Allied Workers (NUAAW)[1] has never been successful. The NUAAW, long threatened and poached upon by the giant TGWU, has recently been swallowed up and become a section of it. Farmworkers have proved to be extremely difficult to organise because of their scattered and often remote places of work, and because of their loyalty to and identification of interest with their farmer employers.[2] Membership of the union has tended to comprise a fifth or less of all farmworkers and the active membership, concentrated in Norfolk, Lincolnshire and East Yorkshire, is a small fraction of the total. As Self and Storing[3] have remarked, the union has always felt itself a poor relation to the NFU which even Labour governments have preferred to consult on agricultural matters, ignoring the union despite its strongly pro-Labour leadership and its sponsored Labour MP, Joan Maynard. The union has been undecided about its true rôle. Should it play the traditional trade union adversarial rôle of trying to extract maximum wages from the employers, or

---

[1] F. D. Mills, 'The National Union of Agricultural Workers', *J. Agric. Econ.*, Vol. xvi, December 1964.

[2] Even the TGWU found it impossible to gain support for 'industrial action' against farmers during the 1984 harvest.

[3] Self and Storing, *op. cit*.

should it ally itself with the NFU in trying to extract maximum support from government? In its indecision it has not played either rôle effectively.

The National Farmers Union

With the farmers' organisation the story is very different. Although there are a number of bodies, including the National Farmers Union of Scotland, the Farmers Union of Wales, the Ulster Farmers Union, and the Country Landowners' Association, the National Farmers Union of England and Wales is by far the strongest. Founded in 1908 as a tenant farmers' association, the NFU first gained influence with government during the expansion of food production in the latter part of the First World War. But it really came to prominence during and after the Second World War with the growth of state involvement in agriculture. Its membership peaked at 210,000 in 1953, since when it has declined with the number of farmers although the proportion who are members remains remarkably high at possibly 80 per cent or more.

The political influence of the NFU became such in the immediate post-war period that it was held up as a model to which all other pressure groups should aspire. Its strength derived from a number of sources. First, it has brooked no rivals; although some dissatisfied Welsh farmers did break away in 1955 to form the Farmers Union of Wales, it has been recognised as the paramount farmers' representatives by the MAFF, which naturally prefers to deal with one body. Secondly, it built up a very strong leadership and central organisation during the presidency (1945-60) of Mr James Turner (later Lord Netherthorpe, Chairman of Fisons) and the general secretaryship of Mr J. K. Knowles. Although the union appears to have a very democratic structure, it is more in form than substance with power being highly concentrated at the top. Turner and Knowles formed a powerful and professional team, a benevolent dictatorship with a reputation in Whitehall for being able to deliver the goods. They built up in Agriculture House (the NFU's Knightsbridge headquarters) a highpowered secretariat of professional administrators, economists and technical specialists who were able to match the men from the Ministry, from the ranks of whom many of them (including the present Deputy Director) have been recruited. These two bureaucratic groups have maintained a monopoly of knowledge of the technical intricacies of agricultural policy which has enabled the technocrats of the NFU to manipulate its members and the civil servants to manipulate their ministers. The complexity of policy has become such that the ordinary NFU member, or even delegate to its central Council, has insufficient knowledge to contribute on many issues – which are thus left to the experts.

Thirdly, the loyalty of its members is retained by the wide range and quality of services which the NFU offers them. Many of these services are administered locally by the County Secretaries who are the paid officials of the Union with the closest contact with individual members. There is virtually no problem a farmer can have for which the County Secretary cannot provide assistance. The main assistance is in dealing with the vast government machinery surrounding the farmer which ranges from matters affecting all small businessmen (like health and safety regulations, employment legislation, planning controls, and taxation) to such specifically agricultural problems as rent disputes, claims for compensation for the compulsory purchase of land for roads, and payments from oil companies or electricity authorities which wish to lay pipelines or cables over farm land. But the major service is to give advice about claiming the multitude of grants and subsidies which are available under national and EC policies. No doubt it is the dairy quota compensation scheme which is occupying much of the time of many County Secretaries at present. In addition, the NFU has its own insurance companies, the NFU Mutual and the Avon, the former of which deals exclusively with members at highly competitive rates and offers policies tailored to the special needs of farmers. In short, the farmer cannot afford *not* to be a member of the NFU.

Its impressive internal strength would not have been sufficient to enable the NFU to come to political prominence had it not been for the large measure of political shrewdness exercised by Turner and his successors, although none of the latter has been of his stature. A key element of Turner's approach was that the NFU should maintain political neutrality. It does not sponsor MPs, as trade unions do, and it strictly avoids political partisanship by maintaining close contacts with interested members of all parties. Its aim is to work with and influence the government of the day of whatever party. It maintains a full-time Parliamentary lobbyist who keeps in touch with MPs of all parties (maybe 100 or more) who have the slightest interest in or influence on agricultural matters. Even closer contacts are maintained with the civil servants of the Ministry by their NFU counterparts. They meet daily on some matter or another and speak each other's language with perfect fluency. The objective of the men from the NFU is always to know precisely what is going on in Whitehall, to anticipate possible shifts in policy or new policies which might affect farmers, and to influence them long before they are put to Parliament or even to ministers. It is much easier to influence policy when it is in formation than when it has become 'firm'. Such is the quality of the briefs on agricultural issues which the NFU provides to MPs that many speeches in the House of Commons are extracted directly from them.

The closeness of the relationship between the Union and government was recognised in the Agriculture Act of 1947 which enshrines a statutory right for farmers' representatives to be consulted by government at the Annual Review of the agricultural industry. Although the NFU is not mentioned specifically in the Act, it has always been accepted as the farmers' main representative and jealously guards its privileged right of access to the Minister and his officials. In the 15 years after World War II, the NFU was able to exploit its strong position under both Labour and Tory governments, aided by the perceived national need for higher agricultural production at a time of food shortages and balance-of-payments problems and by the politicians' fears of offending the farm vote. Indeed, it came closest to achieving the ultimate aim of any pressure group – namely, political power without having to be elected – until the trade unions were allowed to dictate the policies of the 1974 Labour Government. The peak of its influence was probably reached in the 1950s with the passing of the Agriculture Act of 1957, since when some of its weaknesses have become more apparent as general economic conditions have become less favourable to the Union.

The main weakness of the NFU is that it is not a trade union which can dragoon recalcitrant members into its ranks or into following its policies. Membership is voluntary, there is no closed shop, and no means of disciplining members who step out of line. When all was going well, it was easy to maintain unity and gain ready acceptance for decisions made at the top. Since the 1960s, however, the going has not been so easy and the differences of interest within the union have emerged into the open. There are two main sets of conflict – the arable *versus* the livestock producers, and the large *versus* the small and marginal farmers – which tend to run along parallel lines: the livestock farmers, who are mainly small and frequently marginal, feel aggrieved at what they see as the preferential treatment given to the large arable farmers. By the very nature of his job, the small livestock producer is tied to his farm the whole year round and does not have time to participate in the meetings of the major committees which are held in London. The NFU hierarchy is thus inevitably dominated by large arable farmers. This division of interest has resulted in the splitting away of the Welsh farmers and in a number of unsuccessful attempts to set up separate unions for dairy farmers. Since EC entry, the divisions have become more numerous as the sense of injustice felt by pig and poultry farmers, and now by dairy farmers under the quota system, has increased.

New allies

Indeed, EC entry has posed the more general problem for the NFU of watering down its direct influence on agricultural policy. Policy is now determined mainly in Brussels and can be influenced only at second hand through the British Ministry's representations to the EC Commission and in the Council of Ministers, or through the confederation of farmers' organisations in the Community (COPA) which is one of the most vociferous lobbies in Brussels. Coming at a time when the NFU's influence was waning, however, British entry to the EC has benefited the Union's members by allying them with the immensely powerful French and German farm lobbies. Since agriculture has increasingly become a system of processing industrial products through plants and animals, the NFU has also gained powerful allies in the manufacturing industries which supply farming with vast quantities of fuel, chemicals, drugs, medicaments, buildings, plant and equipment. In his second book, *Farming in the Clouds*,[1] Richard Body calls for more investigation into this relationship and promises a third book devoted to it. To put the matter in perspective, it should be mentioned that, apart from feeding-stuffs (many of which are manufactured from non-agricultural or non-domestic products), the variable inputs (excluding labour) of British agriculture cost £3·4 billion (1983). And the capital inputs of plant, machinery, vehicles, buildings

---

[1] R. Body, *Farming in the Clouds*, Temple Smith, London, 1984.

and works cost a further £1·3 billion – making a total of £4·7 billion. Although these inputs may represent only 2-3 per cent of total manufacturing output, they are very important to a number of major companies like ICI and Unilever. Body gives several illustrations of the activities of the ancillary industry lobby – to which might be added the vigorous encouragement, on balance-of-payments grounds, which Lord Netherthorpe as Chairman of Fisons gave to the campaign for agricultural expansion in the late-1960s. As the pension funds and insurance companies have expanded their investment in agriculture, yet another group has come to have a vested interest in the maintenance of farm policy.

In addition to its new supporters in the CBI and the City, the NFU is not short of friends closer to home. Apart from the other farmers' unions, there is the Country Landowners' Association (CLA), which mainly represents the larger farmers and landowners and has become much more politically active and effective in the last decade. And a host of quangos and quasi-government organisations all add their voices to the chorus in favour of subsidisation: Body lists 67 minor organisations 'whose purpose in some way is to extend the empire of protectionism'.[1] He does not, however, mention some more important organisations such as the Meat and Livestock Commission, the Homegrown Cereals Authority, the marketing boards (of which the Milk Marketing Boards are by far the most powerful and jealous in guarding their members' interests), the Farmers Club, the Royal Agricultural Society, and all the county agricultural societies.

The individual strength or effectiveness of any of the component members of the agricultural producers, processors, and suppliers' organisations may be questioned. When they are added together, however, they make a formidable lobby – even without MAFF whose inclusion makes it an unequalled combination.

The Ministry of Agriculture, Fisheries and Food

It may seem strange to include in a lobby the Ministry which is being lobbied and which is meant to look after the interests of the food consumer as well. But the relationship between MAFF and the NFU is so long-standing and so incestuous that the Ministry has long been just another voice for the agricultural interest. Successive Ministers of

---

[1] R. Body, *ibid.*, p. 101.

Agriculture, most of whom have themselves been farmers, have seen their main rôle as speaking for the industry and 'looking after' the farmers. Those, like John Silkin in the Labour Government of the mid-1970s, who were not farmers and initially were not especially kindly disposed towards them soon found it hard to resist the cosy relationship with the NFU – epitomised when, on first becoming Minister, Mr Silkin received a 'Dear John' 'Yours Henry' letter from the President of the NFU (then Henry Plumb) whom he had never met and whose name he did not even know.

The MAFF civil servants not only tend to suffer from the classic bureaucratic inertia which keeps them doing the same thing year after year because they have no incentive to ask why they are doing it and whether it is still necessary; they also have a strongly entrenched personal interest in support policy which directly or indirectly involves the bulk of them. They conform to William Niskanen's[1] model of bureaucracy which assumes that the bureaucrat is human like everyone else and will thus pursue his own self-interest rather than some idealistic concept of the general welfare. In the absence of incentives to economise on the money he spends and the staff he managers, the bureaucrat will strive to increase the manpower of his bureau and maximise its budget. The more money he spends and the more subordinates he has, the larger his prestige among his peers and the deeper his satisfaction. No one can blame him for thus operating rationally within the system as he finds it. If we wish him to act differently, we will have to find ways of rewarding him personally for making savings on money and staff. Any attempt to simplify the system of support, or to reduce or abolish it, will otherwise attract strong opposition and the minimum of co-operation from the vested interests in the Ministry.

The Ministry does, however, have to compete with other spending departments for Treasury funds. And it does meet opposition from the Department of Trade and the Foreign Office which are keen to ensure that agricultural protection does not damage Britain's external trade and the interests of its overseas suppliers. In this respect, the position of MAFF has been easier since Britain joined the EC because much of agricultural spending and farm trade policy is now determined under the CAP – the bureaucrat's ultimate bonanza – which is largely outside MAFF's control but enormously to its benefit.

---

[1] W. A. Niskanen, *op. cit.*

The Agricultural Vote

The avenue to power and prestige for the politician lies in maximising the number of votes he gains. There can be no explanation for the Labour Party passing the 1947 Act and continuing thereafter to take an extremely generous attitude to capitalist farmers other than a belief on its part that votes were to be won. Labour apologists argue that the 1947 Act was part of its general policy of physical controls on production and prices. But did the farm prices have to be so generous? And why was the long-held aim of land nationalisation kept firmly under wraps? According to Self and Storing, the Labour Party courted the farmers vigorously because it believed that a one-in-four swing in its favour of farmers, farmworkers and their wives could win it 19 more seats in 1955.

Any analysis of agriculture's voting strength must be based on the assumption of the 'malleability' of the farm vote, that farmers can be induced to shift their votes from one political party to another in response to different proposals and actions on agricultural policy. Unfortunately, there has not yet been an occasion to test the validity of that assumption because there has been so little difference between the parties. At the time when Labour was being so generous to the farmers, the Conservatives and Liberals were proposing equal or even more generosity. No wonder Labour did not succeed in attracting Conservative farm voters.

Let us nonetheless assume that farmers would shift their allegiance. What would a political party stand to lose by proposing a radically different agricultural policy of the kind outlined in the last Section of this *Hobart Paperback*? At first sight, the political strength of the farming vote in Britain appears negligible since less than 2·5 per cent of the working population (based on the 1981 census) is officially engaged in agriculture – a smaller proportion than in any other major country. Farmers themselves account for only 0·55 per cent of the total electorate and the whole male agricultural population for around 1 per cent. If wives and other adult dependants are added (assuming one adult dependant per male member of the farm population), the maximum voting strength of British agriculture is around 840,000 people, equivalent to about 2 per cent of the total electorate. However, although farming's importance is relatively small in the country as a whole, it is much larger in many rural areas and Parliamentary constituencies. For

1955, Self and Storing identified 110 'agricultural seats' in which 15 per cent or more of the total male employees aged 15 and over were in agricultural employment. By the late-1960s the number had fallen to 74, of which 52 were Conservative, 13 Labour, and nine Liberal.[1] The most recent research[2] shows that by 1981 there remained only 10 constituencies in Great Britain where more than 15 per cent of the working population was employed in agriculture. A further 29 constituencies had more than 10 per cent of the working population in this sector.

Although the number of agricultural seats in Parliament has declined drastically over the past 25 years, their distribution between the parties has changed very little. Based on the results of the 1959, 1964 and 1966 General Elections, my own earlier work[3] showed that, on average, 72 per cent of them were held by the Conservatives, representing 19 per cent of their total seats in Parliament. Only 3·6 per cent of Labour seats were agricultural, whilst for the Liberals the figure was 82 per cent. Following the 1983 General Election, the Conservatives hold 76·9 per cent of all agricultural seats (7·5 per cent of their total seats); the Labour Party now has only one such seat, Carmarthen; and the Liberal/SDP Alliance share has fallen to 30·4 per cent (seven out of its 23 seats).

Voting significance

The identification of agricultural seats tells us where farmers are likely to be most vociferous, but it does not necessarily tell us much about farming's ability to influence the result of a General Election. The voting *significance* of people in farming depends on factors other than simply their numbers. In any seat, the agricultural vote of electoral significance consists of that portion of the agricultural electorate which is likely both to cast its vote and to be influenced to shift its political allegiance in response to the proposals and actions of the parties on agricultural policy. Since it is doubtful that the voting habits of farmworkers are much influenced by either government agricultural policies or the affiliation of the NUAAW to the Labour Party, the agricultural vote which corresponds most closely to these conditions is that of the farmers.

---

[1] R. W. Howarth, in *Political Studies* (1969), *op. cit.*
[2] J. D. Green, *op. cit.*
[3] *Loc. cit.*

Let us assume that the 'farmers' electorate' in a particular seat is twice the number of active and retired farmers and farm managers, to make allowance for their wives and adult dependants. The farmers' *vote* will, of course, be smaller since only a proportion of the farmers' electorate will actually use its vote; let us also assume, therefore, that their rate of turnout is the same as the average of the *whole* electorate in a particular constituency. To discover those constituencies where the farmers' vote could affect the result of an election, the analysis must be taken a step further to include an assessment of the numerical size of the farmers' vote in relation to the size of the present majority in each constituency. The key to the farmers' political strength is the size of parliamentary majorities, not only the number of farmers. *The agriculturally significant seat is where the farmers' vote is equivalent to half the current simple majority plus one.*

We can thus envisage a seat with a total electorate of 60,000, a majority of 100, and a farming vote of 51. In such a seat, the result at the last election can be changed at the next if all the farming voters can be persuaded to switch their allegiance – assuming all of them voted for the winning party at the last election and that all other voters stay faithful to their previous allegiance at the next. This definition indicates the *maximum* number of seats in which the farmers' vote alone could theoretically affect the result, because it includes those seats where every farming voter may be required to switch from one party to another – an unlikely event. The larger the proportion of the farming vote necessary to exceed half the majority, the less will be the significance of the seat.

At the time of the 1970 General Election there were 60 agricultural seats which were politically significant, of which the Conservatives had 40, Labour 11, and the Liberals nine. In more than half of them, however, 40 per cent or more of the farmers' votes would have had to change allegiance to affect the result – which is a big swing allowing for the unrealistic assumption that all farming voters voted for the winning party at the previous election. Accepting that most farmers vote Tory, the maximum likely swing in the farmers' vote away from the Tories in Tory-held constituencies would be about 40 per cent. In 1970, such a swing could have lost the Conservative Party 16 of their agricultural seats – but a swing of 25 per cent would have lost them only six. The maximum likely swing in Labour-held seats would be about 20 per cent, assuming that a far smaller proportion of farmers vote Labour than Tory. In the event of such a swing, Labour stood to lose only three of its agricultural seats in 1970.

Present political significance

The number of politically significant agricultural seats remaining after the 1983 General Election has dropped to a mere 12. Of these, only five are now Conservative. The Alliance also holds five, and Labour and Plaid Cymru hold one each. In eight of those 12 seats, 50 per cent or more of the farmers' votes would have to change their allegiance to affect the result at the next Election, a most unlikely occurrence. A more realistic swing of 25 per cent would affect the result in only four. The Conservatives would lose two (Banff and Buchan, and Clwyd South West); the Alliance would lose one (Montgomery); and Labour would lose Carmarthen. Montgomery and Carmarthen are especially vulnerable and could change hands with less than a 10 per cent swing in the farmers' vote.

Widening the net from the agricultural to *all* the British Parliamentary constituencies, a further 14 seats (of which the Conservatives hold five) may be agriculturally significant. However, much of the 'agricultural vote' is not truly agricultural in these seats. In only two of them does the farmers' vote exceed 2·5 per cent of the total vote, which indicates that they are predominantly urban constituencies where the agricultural population mainly comprises people like market gardeners and smallholders whom the Census data do not eliminate from the 'agricultural' groups. Nevertheless, they should be included in the analysis to find the maximum possible electoral strength of agriculture. Assuming again that the maximum swing likely to occur is 40 per cent in Conservative-held seats and 20 per cent in Labour-held ones, the Conservatives might lose a further two seats, the Alliance one and Labour none.

There is thus no danger whatever to the Labour Party from losing whatever agricultural vote it has. The Liberals, however, remain vulnerable in the Celtic fringes, which explains why they are strongly opposed to liberalising agriculture. In the past, the Conservatives probably have had some reason to fear offending the farmers' vote. As the pattern of farming has changed and the number of farmers has declined, however, the electoral danger to them from farmers has receded to an absolute maximum of 10 seats, and a realistic maximum of four or five. There no longer appears to be any electoral obstacle to the Tories adopting a radical stance on farm policy.

Agriculture in Parliament

For a Conservative Prime Minister contemplating change, a bigger obstacle than the farming voters is the number of her colleagues who are themselves farmers or who come from constituencies where farmers dominate the local party. The farmer MPs and Cabinet Ministers have a vested interest in the present support system, and those from farmer-dominated constituencies are likely to prefer a quiet life to the violent antagonism they would arouse by advocating more liberal policies.

The occupational representation of agriculture in Parliament in the post-war period has been remarkable. In the inter-war years there were on average only 10 MPs whose principal occupation was farmer or stockbreeder. There were 16 in 1945, 25 in 1950 and 33 in 1951. At the 1955 General Election, as Butler[1] points out, 36 farmers were elected. According to Beynon and Harrison,[2] 79 MPs with a farming background were elected at the 1959 General Election, of whom 56 were farmers, 16 landowners, and seven 'other'. Forty-eight of the farmers were Conservative and eight Labour; all of the landowners were Conservative, as were six out of the seven 'other'. These figures are somewhat higher than those of Butler,[3] who records that 38 Conservative and three Labour farmers were elected in 1959.

A large number of the big crop of MPs with a farming background identified by Beynon and Harrison lost their seats in 1964 and 1966. After the 1966 General Election, only 38 remained of the 79. Of the farmers, 24 retained their seats and 20 of them were Conservatives. Nine of the 16 landowners were returned in 1966. There must, however, have been a number of farming newcomers in the Parliaments of 1964 and 1966 to replace some of the 1959 entrants who lost their seats. In 1964, farming was the 'first or formative' occupation of 52 Conservative candidates (35 elected), of eight Labour candidates (two

---

[1] D. E. Butler, *The British General Election of 1955*, Macmillan, London, 1960, p. 43.

[2] V. H. Beynon and J. E. Harrison, *The Political Significance of the British Agricultural Vote*, Report No. 134, Department of Economics (Agricultural Economics), University of Exeter, 1962, pp. 39–40.

[3] D. E. Butler and R. Rose, *The British General Election of 1959*, Macmillan, London, 1960, p. 127.

TABLE XVII

NUMBERS OF MPs WITH AGRICULTURAL AFFILIATIONS ELECTED AT THE 1983 GENERAL ELECTION

	Conservative	Labour	Liberal/ Alliance	Total
Active or Hobby Farmers	29	1	1	31
Landowners	7	1	—	8
Special Interest*	17	2	1	20
Total	53	4	2	59

*Includes the directors of agricultural firms and all MPs who indicate in *Dod's Parliamentary Companion* that they are members of the NFU.

elected), and of 16 Liberal candidates (one elected).[1] The numbers were similar after the 1966 Election. According to *The Times*,[2] farming and landowning were the main occupations of 38 Conservative, four Labour and three Liberal Members.

At the 1970 General Election, the number of agriculturally affiliated Members rose to 90 – 70 Conservative, 10 Labour, and one Liberal. By 1979, MPs with agricultural backgrounds, either farmers or landowners, had fallen to 30 – of which 25 were Conservative, two were Labour, and three were Liberal. But their number has since risen again. Leaving aside the House of Lords where nearly all the hereditary and several Life Peers are landowners, the House of Commons contains 59 members who are closely connected with agriculture – which is approximately 9 per cent of all MPs. Table XVII breaks them down according to their party and whether they are farmers, landowners or merely have a special interest in agriculture. Not surprisingly, all but a few of the agricultural members are Conservatives. The close connection between the Tory Party and farmers remains firmly intact.

Undoubtedly, farmers benefit from the organised support of a surprisingly large number of MPs with agricultural backgrounds or an interest in the industry. From the agriculturally-minded members

---

[1] D. E. Butler and Anthony King, *The British General Election of 1964*, Macmillan, London, 1965, pp. 234, 235.
[2] *The Times Guide to the House of Commons 1966*, London, 1966, p. 242.

can be expected a strong commitment to the perceived interests of agriculture as a whole, especially by MPs who represent the rural constituencies. Indeed, of the 59 MPs with agricultural affiliations, 12 represent 'agricultural constituencies' and can be expected strongly to support the views of their constituents on agricultural policy. However, the most influential Parliamentary representatives are those who sit in the Cabinet, and *Dod's Parliamentary Companion* shows no fewer than 10 Cabinet Ministers having close connections with agriculture. This small but powerful élite may be the strongest link in the chain of agricultural interest groups. As a recent article in *The Times* remarked, '. . . the Prime Minister may face determined opposition within the Cabinet if she presses ahead with an examination of the privileges of farmers'.[1]

It is reasonable to assume that the agricultural lobby will strongly resist any attempt to change the basic principles or volume of farm support. However, with a large Parliamentary majority and a reputation for being able to dominate her Cabinet, the present Prime Minister is in a stronger position to achieve agricultural reform than any previous Conservative holder of that office. As for the principal opposition party, Labour, it has no justification – electoral, Parliamentary, or ideological – for giving special treatment to farmers.

---

[1] *The Times*, 8 March 1984.

APPENDIX TO SECTION VI

List of Main Sources

Given the importance of the subject, relatively little has been written on the politics of agriculture, and very little since the 1960s. There are considerable gaps in the literature which require to be filled. The main material for this section has been gathered from the following sources:

Allen, G. R., 'The National Farmers Union as a Pressure Group', *Contemporary Review*, May and June, 1959.

Beynon, V. H. and Harrison, J. E., *The Political Significance of the British Agricultural Vote*, Report No. 134, Dept. of Econ. (Ag. Econ.), University of Exeter, 1962.

Kirk, J. H., 'The Economic Activities of the Ministry of Agriculture, Fisheries and Food', *J. Agric. Econ.*, Vol. xvi, December 1964.

Mills, F. D., 'The National Union of Agricultural Workers', *J. Agric. Econ.*, Vol. xvi, December 1964.

Pennock, J. R., 'The Political Power of British Agriculture', *Political Studies*, October 1959.

Howarth, R. W., 'The Political Strength of British Agriculture', *Political Studies*, December 1969.

Donaldson, J. G. S. & F., and Barber, D., *Farming in Britain Today*, Allen Lane, The Penguin Press, 1969.

Mackintosh, J. P., 'The Problems of Agricultural Politics', *J. Agric. Econ.*, Vol. xxi, No. 1, January 1970.

Howarth, R. W., 'What's Your Vote Worth?', *Farmers Weekly*, 22 May 1970.

Green, J. D., *An Analysis of the Political Strength of British Agriculture with Special Reference to the 1983 General Election*, unpublished dissertation, Dept. of Agriculture, University College of North Wales, Bangor, June 1984.

Phillips, M. C., *An Analysis of the National Farmers Union as a Political Pressure Group*, unpublished dissertation, Dept. of Agriculture, UCNW, Bangor, June 1984.

VII. Future Policy

Summary of the Argument so far

This *Hobart Paperback* has argued that agricultural support policy during World War II and the immediate post-war period succeeded in its original aim of rapidly expanding the supply of temperate foodstuffs, aided by major technical advances, but has been totally unsuited to the changed circumstances from the mid-1950s onwards. The forces of supply and demand have led to a propensity to persistent oversupply of food which has been aggravated by price support and input subsidy policies. The result has been a much increased burden of costs borne by taxpayers and – under the CAP particularly – by consumers. The burden on consumers has recently been estimated to represent a 4·5 per cent tax on gross incomes in Britain which is spread regressively across income groups.[1] The support may have protected many farmers from the perils of 'adjustment' and reduced outward migration.

Contrary to their beliefs and those of the NFU, however, most farmers – especially tenants – have not benefited much from support; it has not raised their relative income position as a group. A high proportion of the *gross* benefits they have received in price and input support has been dissipated in high land values, rents, and interest payments, which have in turn provided a barrier to new entrants to the industry. Indeed, it has been argued that there is a 'natural rate of disparity', or a natural 'low income', in agriculture in Britain and similar advanced countries which no amount of support will alter. So long as there are large numbers of people willing to farm for apparently low relative cash incomes, the 'farm problem' will not disappear because it is not a problem. And even if there were an issue of 'trapped resources' forcing people to accept low incomes in agriculture, the position is not as bad as it is portrayed because farmers have other exploited and potential sources of income. Farm policy has also had adverse external consequences: its example is an obstacle to progress

---

[1] Dilnot and Morris, *op. cit.*

in reducing other sources of misallocation and other rigidities in the national economy; it has damaging environmental effects; it retards the liberalisation of international trade; and it damages the progress of agriculture in developing countries.

In short, while support policy may have helped maintain the number of farmers, it has not achieved its objective of raising farmers' relative incomes – which was probably neither desirable nor attainable in the first place. Furthermore, it has imposed a heavy and highly inequitable burden on others. If support policy was ever politically necessary, it is no longer so today.

It follows that current agricultural policy ought to be fundamentally changed and support reduced (if not abolished outright). Two main questions arise: 'Where do we go from here?' and 'How do we get there?'. A start at answering them can be made by outlining what the functions of government, if any, should be in agriculture.

Food Security

Strict disciples of Adam Smith would argue that there is no rôle whatever for government in agriculture. I would not go to that extreme. Any government which does not safeguard the food supply of its people will not last long in a democratic society. By 'safeguarding' I mean two things. First, there should be adequate food to maintain minimum nutritional standards for everyone if an emergency, war, or natural disaster affected overseas food supplies. Secondly, plentiful supplies should always be available in normal times to maintain roughly the present diet (or whatever diet the people demand) at current, or preferably lower, prices.

To take the second criterion first, it is beyond dispute that support policy has fed the people plentifully. However, a free market would do the same at generally lower prices and without the adverse consequences. As for international emergencies, government has a clear responsibility. At present, the only type of emergency which could foreseeably cut off or drastically curtail Britain's overseas food supplies for a long period is a war – either in Europe or involving North America or Australasia. It could come about only if one or more of the countries in those regions were attacked; and the only possible attackers are Russia and, remotely, China. Since both are nuclear powers any ensuing war would inevitably be nuclear, with such devastation that any possible agricultural policy we can think of would be irrelevant. The old strategic argument about safeguarding Britain's food supply against the eventuality of attacks on shipping which carry its imports has become as obsolete as the Spitfire.

In the longer run, there is a possibility that the freak market conditions of the early-1970s will recur; that the Third World as a whole will find the magic key to economic success and become a dominant international buyer of foodstuffs, thereby threatening Britain's supplies; or that, as a result of some international dispute, agricultural politics will come into play using food exports as a bargaining counter (America tried that game against Russia in the wake of the Afghanistan invasion but soon stopped under pressure from its own farmers). However, with multiple sources of supply for most of the commodities it imports, it is difficult to see how Britain could be hurt, and by whom.

What alone is predictable is that the unpredictable will occur. Government has a duty to secure Britain's minimum food requirements by holding sufficient stocks in excess of national production for whatever period is deemed necessary to bridge the time gap until it can be expanded. The latter demands that production capacity, land and non-imported inputs should be readily available for expansion. The emergency argument does not require the maintenance of an artificially big agriculture which supplies all Britain's requirements all the time, but rather, as Professor Gerald Wibberley puts it, an agriculture 'with sufficient, well-equipped and knowledgeable farmers in it who could quickly turn over poor and unused land to increased food production'.[1]

There is another side to the coin. A country close to self-sufficiency is vulnerable to domestic natural disasters – droughts, floods, crop and animal diseases – and to being held to ransom by militant farmers. If it lacks a developed importing system – port and storage facilities and import agencies with experience, equipment, know-how and trading links – its food supply it likely to be more vulnerable than that of a country which is much less self-sufficient. This argument was advanced by Christopher Ritson in response to the 'world food crisis' of 1974-75:

'When considering levels of domestic production in excess of those required to meet the minimum nutritional needs of the population, the case for

---

[1] G. Wibberley, 'The Diversion of Land', in CEAS/CAS, *op. cit.*, p. 32.

increasing agricultural self-sufficiency to increase the security of food supplies rapidly evaporates'.[1]

It is impossible to be certain that Britain would produce enough food to meet its minimum requirements if agricultural price support were abolished and world prices were permitted to prevail. However, an educated 'guesstimate' is that it probably would. The introduction of world prices would reduce 1982 agricultural prices by around 25 per cent on average. Using Christopher Johnson's[2] estimate of a 0·6 price elasticity of supply for British agriculture, a 25 per cent price fall would lead to a 15 per cent reduction in output. A 15 per cent reduction in Britain's current 80 per cent self-sufficiency would still leave it about two-thirds self-sufficient in temperate goods. If two-thirds of current consumption in Britain was spread evenly among the population, there is little doubt the latter could survive adequately – though on an admittedly less varied diet. Nutritionists generally would agree that most of us eat at least a third too much for the good of our health.

Provisions Against 'Dumping'

There is also an argument for a country to have a policy against the dumping of agricultural products, despite the benefits of unilateral free trade. This is because supplies which are dumped are normally sporadic and unreliable (although the dumped surpluses of the CAP have been going on for a remarkably long time); they can undermine home producers to the extent of putting them out of business and thereby endangering food security. Dumping, however, needs to be clearly and carefully defined so that it is not used by producer lobbies as a pretext to press for the banning of all competition from cheaper imports.

The definition offered by Seldon and Pennance[3] equates dumping with price discrimination – 'selling goods abroad at prices below those in the exporter's home market'. This is unsatisfactory because price discrimination *per se* is a perfectly legitimate and common commercial practice for exploiting differences in elasticities of demand in different

---

[1] C. Ritson, *Self-Sufficiency and Food Security*, CAS Paper 8, University of Reading, March 1980.
[2] C. Johnson, CEAS/CAS, *op. cit.*, p. 64.
[3] A. Seldon and F. G. Pennance, *Everyman's Dictionary of Economics*, J. M. Dent, London, 1965, p. 131.

markets. 'Selling goods abroad at prices below their costs of production' would be a better definition because it implies an overt or disguised export subsidy. An import duty, calculated to offset the subsidy, could be imposed to counter this practice. It would not, however, be feasible to tailor an individual duty to each of a large variety of subsidies. A better idea would be to establish a threshold price[1] for each commodity based on the supply price from the most efficient, least subsidised major exporting country – which would normally be the country with the largest natural comparative advantage (for example, milk products from New Zealand, cereals from North America, and beef from Argentina). Where producers in the most efficient exporting countries were nonetheless receiving some subsidy which reduced their production costs and selling prices, an offsetting adjustment would have to be made to the threshold price. Threshold prices would be set (weekly or monthly) for a small number of key commodities and offers below them would be subjected to a variable import levy.

Although I am reluctant to propose such a bureaucratic mechanism (which is now a part of the CAP and was part of the former British policy), there seems to be no alternative in a world where so many countries subsidise their farmers and their agricultural exports. It would be particularly foolish to become dependent on subsidised exports from Eastern bloc countries. Although Britain should and could return to a much freer market in agriculture, it is unrealistic to assume that all other countries would follow suit immediately – although it would be equally desirable for them. The experience of Denmark in the late-1950s and the 1960s demonstrated it was impossible to have a largely free-market agricultural policy while attempting to export two-thirds of total output in a protectionist world; and it led that country to introduce assistance for her farmers on a minor scale which included levies on dumped imports of feedgrains. With the advent of the CAP in which one of her major customers, West Germany, was a founder participant and another, Britain, was a prospective one, Denmark was forced to submit to the old maxim: 'If you can't beat 'em, join 'em'.

---

[1] By which is meant a minimum import price.

Other Minor Functions

Government should retain a number of uncontroversial activities in agriculture which were referred to earlier as 'technical policies' and 'research, education, and extension'. Few people would wish to dispense with enforced standards for public health and hygiene and animal health and welfare, or with controls over the use of chemicals and drugs. The collection of statistics is also necessary, but ought not to be carried to the present extremes.

It has already been argued that state-financed advice to farmers and much research into temperate agriculture is unnecessary. Agricultural research can be carried out or financed perfectly satisfactorily by commercial companies; and speculative research in universities ought to be financed from their normal budgets. There is a humanitarian argument for assisting research and advisory work in developing countries, but this is probably best carried out through voluntarily-funded agencies.

Agenda for Immediate Action

In the light of these limited functions for the state, a number of proposals for future policy may be made. The time-scale for their implementation, however, is a matter for debate. Richard Body appears to want to abolish the whole support system almost overnight, which is easier said than done. Immediate abolition, regardless of its political feasibility, could have such major repercussions on the whole economy as to be undesirable in any event. The problem is that high price support has encouraged farmers to expand their output (urged on by Ministers of Agriculture, up to and including Mr Peter Walker less than two years ago) and to make costly new investments in land, buildings, capital equipment and livestock – in many cases with the aid of expensive borrowed money. A rapid about-turn in policy would cause severe financial difficulties not only for farmers but for those who supply them, for the land market and thus the security of all agricultural borrowers and lenders, and for the rural economy in general. Production periods for agricultural products, and the payback period for agricultural investments, are so long that it requires time to change direction. A period of five to 10 years would therefore seem desirable to phase out price support and allow for an orderly, but painful, adjustment of the structure, output and inputs of farms.

In the short run, however, there is a lot which can be done to reduce public spending on and state involvement in agriculture, particularly that part which is the sole prerogative of the British government and is not compulsory under the CAP.

(1) *Marketing boards*

An early *Hobart Paper*[1] strongly criticised producer-controlled marketing boards – many of which have been monopolies – and which are a legacy of the 1930s when conditions were very different. These criticisms remain valid, although the only remaining boards are for potatoes, wool, and milk. The Potato Marketing Board is currently under heavy fire from dissident producers and will probably disappear. The Wool Marketing Board deals with a very minor product and is responsible for administering a small subsidy (an estimated £3·2 million in 1983-84). But the Milk Marketing Boards received an overwhelming vote of confidence from producers in a 1979 referendum and continue to control the market for the single biggest product of UK agriculture (with an output currently worth £2,500 million). By operating a discriminating monopoly, the Milk Boards have long been responsible for maintaining very high prices for liquid milk in Britain, although the CAP must now bear equal responsibility. The Boards remain committed to restricting consumer choice – witness their long battle to prevent the import of small quantities of UHT milk – and should be disbanded and re-formed, if that is what producers want, as proper co-operatives shorn of their powers to manipulate the milk market and to control unwilling producers. They would be able to continue their useful promotional, advisory, and manufacturing activities in a competitive environment.

(2) *Support for capital and other improvements*

Under this heading come a whole host of different grants and subsidies, some of which are partly funded from the EAGGF. They include the Agriculture and Horticulture Development Scheme (an EC scheme) and the Agriculture and Horticulture Grant Scheme (a national one) which are designed to encourage investment in buildings, field drainage, water supply and other 'improvements'. They have encouraged *wasteful* investment (in that much has been used to produce uneconomic

---

[1] G. Hallett and P. G. James, *Farming for Consumers*, Hobart Paper 22, IEA, London, 1963.

output), and have also been responsible for much environmental damage (draining marshes, etc.).[1] In total the schemes cost £197·2 million in 1983-84 (estimated)[2] and there is no reason why they could not be abolished forthwith and the expenditure saved. Obviously, farmers who were part way through an agreed scheme would have to be subsidised until the end of their plan, which could last up to six years. The saving in the administration of about 70,000 projects a year would also be useful.

(3) *Support for agriculture in special areas*
Support for farming in 'less favoured areas' comprises mainly higher rates of grant paid under capital grant schemes[3] and the special 'hill livestock compensatory allowances' (another EC scheme which has replaced the very similar hill cow and hill sheep subsidies). It also includes small sums disbursed under special schemes for Northern Ireland and for Scottish crofters.

Support under these headings amounts to around £120 million a year divided among 52,000 recipients[4] – which is not a lot to pay for maintaining the population and infrastructure of these areas in Wales, the Pennines, the Highlands and Islands. But it could be spent more wisely. Paying the farmer per head of cattle and sheep simply encourages over-stocking by 'farming for the subsidy' and thereby adds a little bit more to over-production while encouraging over-grazing which detracts from the amenity of the landscape. Along with grants for 'improving' hill land, the 'headage' payments have led to the despoliation of much of the rugged landscape. If there is an argument for subsidising farmers in these areas, it is for subsidising the people and not the inputs or outputs of their farms. It is essential that any grants should go to individuals (and should not be marketable), otherwise their value will be capitalised into land values again and they will

---

[1] Grants for reclaiming marshes and grassland were withdrawn in December 1983 following protests from environmentalists. But there is still a 50 per cent grant for ploughing or re-seeding moorland.

[2] Cmnd. 9137, *op. cit.*, p. 46.

[3] Hedge and wall maintenance 60 per cent (instead of 32·5 per cent); field drainage 60 per cent (30 per cent); roads 40 per cent (20 per cent).

[4] An EC agreement of March 1984 designated a further 1 million hectares divided between England, Wales and Scotland as 'less favoured', adding another 28,000 farmers to those already eligible for the special grants at an estimated cost of £10 million a year.

defeat their object. The land could then be used in response to undistorted market requirements, giving a much better mixture of farming, forestry, recreation and tourism. Assistance could become part of a countryside policy without an agricultural bureaucracy. It is beyond the scope of this *Hobart Paperback* to define a countryside policy, which requires much further discussion. But the danger of it becoming simply another way of handing out large sums to farmers, which could again be wasteful and inequitable, should not be overlooked.

(4) *Public expenditure on research and development*

There is little justification for this item, which has spawned numerous quangos. Adding the current annual costs of research commissioned by MAFF (mainly through the Agricultural and Food Research Council) at £59 million, the salaries and travelling expenses of the 4,790 members of the Agricultural Development and Advisory Service at £61 million, and the costs of other 'educational, advisory, research and development services' at £75 million, a total potential saving of £195 million is found.[1] A further £14 million could be saved over the next five years by withdrawing public funding for 'Food from Britain', the new food promotion quango which farmers themselves could finance – perhaps by a levy organised by the NFU.

(5) *National price guarantees*

Wool and potatoes are the only commodities still covered by national price guarantees – through their respective marketing boards. Abolition of these guarantees, which should be phased out over five years, would save a further £14 million based on current expenditure.

(6) *The Forestry Commission*

Although forestry is not within the ambit of this *Hobart Paperback*, like agriculture it has its own tax privileges and its own quango, the Forestry Commission, which was set up during the First World War to expand timber production for strategic purposes. There is no longer any good reason why the state should be involved in raising trees and the process of privatisation begun in the Forestry Act of 1981 should

---

[1] Figures from: *Supply Estimates 1982-83*, HMSO, London, 1982, and *The Government's Expenditure Plans 1984/85 to 1986/87*, Cmnd. 9143-II, HM Treasury, February 1984.

be accelerated and brought to a speedy conclusion.[1] After allowing for sales of land, plantations and buildings, the net cost to the taxpayer is £41·9 million (1982-83).

Adding together the potential savings in public expenditure from items 2, 4, 5, and 6 produces a total of £462 million. Though a not insignificant sum of money, it is a mere fleabite compared with the elephantine costs of price support under the CAP. Major changes in farm policy can therefore be achieved only by tackling the thorny problem of the CAP.

Britain and the CAP: the Options

Britain has a number of options towards the CAP:

(1) *To submit gracefully*

We could stop protesting about the CAP, stop complaining about our budget contribution (which even at £1·5 or £2 billion a year might be regarded as a modest membership fee for the European Community club), and become 'good Europeans' in the eyes of our partners by allowing the majority to have their way and by increasing our contributions whenever asked to finance the ever-mounting costs of the CAP. We could also provide some modest savings for consumers and the Community budget by keeping the green pound in line with the real pound, instead of running a positive MCA which has, in the past couple of years, given us the second highest farm product prices in the Community after West Germany.

(2) *To press for 'reform' of the CAP*

We could go on pressing for changes to the CAP, which might turn out to be either fundamental or tinkering. Fundamental change would entail gradually lowering EC prices to world market levels and directing expenditure to assist the structural shifts which would ensue – compensating, removing, re-training, and pensioning-off farmers. Tinkering or cosmetic change would not attack the root problems of the CAP but would be a temporary expedient causing further rigidities, inequities, bureaucracy, and scope for fraud (quotas and super-levies are examples).

---

[1] A critique of the Forestry Commission and of forestry policy can be found in R. Miller, *State Forestry for the Axe*, Hobart Paper 91, IEA, London, 1981.

(3) *To press for the scrapping of the CAP*
We could try to persuade our fellow EC members that there was no chance of ever devising a common agricultural policy to suit 10 countries, let alone 12 with the arrival of Spain and Portugal. Everyone should therefore return to national policies whilst trying to limit the barriers to agricultural trade within the EC.

(4) *To withdraw from the CAP unilaterally*
After pressing for major reform for so long without success and after failing to obtain agreement on option (3), Britain could simply announce that it was no longer going to comply with the provisions of the CAP or contribute to its funding and would revert to its own national farm policy.

(5) *To allow the CAP to die a natural death*
We could refuse to agree to additional 'resources' being made available for the Community (that is, for the CAP), thus provoking an inevitable collapse of the policy; or we could wait until Spain and Portugal become integrated into the CAP when the additional costs could make France a net contributor and much less enthusiastic about the policy.

The Decline of the CAP

In view of the criticisms of price support in general and the CAP's methods of achieving it in particular, the first option of submitting gracefully is not to be recommended. Fundamental reform to a free internal and external market with common structural policies – or anything remotely approaching that ideal – is out of the question, as all experience has shown. It is now over 30 years since common agricultural markets in Western Europe were first discussed, over 20 years since the beginnings of the CAP, and over 15 years since the Mansholt Plan (1968) proposed fundamental reform of the CAP. Dr Sicco Mansholt, the father of the CAP, as the first Commissioner for Agriculture, soon realised that his brainchild, for which there had initially been such high hopes, was a monster with a voracious appetite for other people's money and was running out of control. In 1968, when the CAP had come into full operation among the Six, he went back to the drawing board and produced proposals for radical change which have formed the basis for numerous subsequent plans. Neither the Mansholt Plan nor any of its successors has yet been adopted; and none is likely to be in the future.

Even the basic principle of a free agricultural market within the EC has been steadily eroded. Since the late-1960s, monetary problems and the advent of green money and MCAs have ruined any chance of unobstructed internal trade in farm products. Indeed, the paperwork for internal trade is said to be larger than for external. Non-tariff barriers have grown rather than diminished. And manipulation by governments of their green currencies has made a mockery of the principle of common prices.

Not Quotas

We now have milk quotas (and the prospect of cereal quotas) following the failure of the 'co-responsibility levies' introduced in 1977 under which farmers paid a flat-rate tax on their production equivalent to a small price cut. Those levies proved ineffective in controlling the mounting milk surplus. Milk quotas, which are the latest palliative, have already proved to be the administrative nightmare their critics had predicted. The scheme was introduced in June 1984 with insufficient preparation and has thrown the dairy sector into a state of confusion and panic which is reverberating throughout the farming industry.

All British milk producers have been allotted a quota based on their output in 1983 less 9 per cent which is the British Government's interpretation of the EC agreement to limit milk output to its 1981 volume plus 2 per cent. Production in excess of a quota will be subjected to a 'super levy'. The amount of super levy the Milk Marketing Boards (through which the scheme is administered) will pay,[1] will be the current target price for milk (17·47 pence a litre) on every litre they receive above the global quota. However, the rate at which the levy is charged on *individual* producers who exceed *their* quotas will probably be less than the full super levy because some producers will have under-produced and others will have gone out of business.

The whole scheme is unfair between countries, not all of which have excess production, and between dairy farmers as it is operated in

---

[1] The super levy will be paid to the EC's agricultural fund via the British Ministry of Agriculture.

Britain. It is a harsh blow to those who had been led up the garden path by the Ministry of Agriculture into expensive schemes to expand their output. It seems particularly unjust that producer-retailers, who may have local customers for their entire output, are also to be allotted quotas. As a result of the problems the system is causing, a special hardship scheme and a compensation scheme have had to be introduced under which MAFF will pay £650 per cow to farmers who choose to get out of milk permanently. However, since the quota is tied to the farm, it is distorting land values between farms with and without quotas. It is also causing friction between landlords and tenants over whose is the quota and who should receive the compensation. Even if the quotas were to become saleable, it is unfair that new producers should have to pay to enter the milk market and that quota owners should be given a free gift by the state. Inevitably, the pattern of production will be ossified and endless wrangles and bitterness between individuals will ensue. There could also develop a black market in milk in which farmers who have filled their quota sell their excess production at low prices to fellow farmers or consumers.

On every occasion when the simplest, most effective and, in the long run, fairest answer to agricultural over-production has been a (phased) reduction in the price of a product, governments have avoided it. They have fudged the issue by bringing in convoluted schemes. The agricultural history of France, Germany, Britain and the USA is littered with them; and the CAP is merely the latest offender in a long line. Yet there is every prospect that the CAP will extend such bureaucratic folly to more and more products.

Differences in Philosophy

Not only do history and current experience show that the CAP cannot be sensibly reformed; differences in philosophy between the members of the EC are such that it never will be. The CAP was created as part of a bargain between the French and the Germans. The EC was to promote free industrial trade to please Germany and to have a common agricultural policy to please France. Germany was sceptical about the CAP because it feared internal political problems from exposing its more heavily-protected farmers to competition. The CAP was therefore moulded in the French tradition (although it adopted mechanisms also used in German farm policy) which has its foundations in 18th- and

19th-century political economy – in particular in the writings of François Quesnay, the founder of the Physiocratic School. Walter Eltis has recently contended[1] that the French are still Physiocrats pursuing their basic tenet that high food prices are the route to economic progress in direct conflict with the British tradition of free trade which led to the abolition of the Corn Laws.[2]

The French attitude has been expounded by Michel Petit[3] who maintains that there is a distinctive French School of agricultural economics and that it has moved from 'rural fundamentalism' to Marxism, neither of which is sympathetic to liberalising agricultural trade. Whether philosophers and economists will ever reconcile the differences between conflicting theories of political economy is uncertain. What is fairly certain, however, is that French farmers, who have a long tradition of '*action directe*', will never allow their politicians to express openly a belief in Ricardian principles of international comparative advantage and the division of labour. It follows that no French government would be likely to agree in the foreseeable future to either a fundamental reform of the CAP or to the abolition of support.

The CAP That Doesn't Fit

Britain is thus left with Options 4 and 5: to withdraw unilaterally from the CAP, or to deny it further resources and be patient while it dies a natural death. It would obviously be *economically* desirable for Britain to leave the CAP as soon as possible provided it could remain inside the Community. It would also be very much to the advantage of the Community as a whole to scrap the CAP while continuing with other aspects of a common market. Most of the other member states do not see it that way, however; and nor will Spain and Portugal. The CAP has been the price of the ticket for Britain's belated entry to the EC, a price now paid in full after 10 years of membership.

---

[1] W. Eltis, 'EEC Agricultural Prices: Are the French Still Physiocrats?', *Economic Affairs*, Vol. 4, No. 2, January 1984.

[2] The apparently irreconcilable differences in philosophy towards agriculture among EC member states have also been discussed in a Trade Policy Research Centre publication which contains contributions from France, Germany, and Holland: H. Priebe, D. Bergmann and J. Horring, *Fields of Conflict in European Farm Policy*, Trade Policy Research Centre, London, 1972.

[3] M. Petit, 'Is there a French School of Agricultural Economics?', *Journal of Agricultural Economics*, Vol. xxxiii, No. 3, September 1982.

As *The Economist*[1] has remarked: 'It would be difficult to devise a farm policy less suited to British circumstances than the CAP'. In Britain, agriculture accounts for a smaller share of the economy than in any other EC country and its farmers now have the least political weight. Britain is the largest food importer, and the smallest food exporter in relation to its total exports. It has had a long tradition of importing food from the most economic sources, and its housewives have been used to a wide variety of food at the lowest prices. In contrast, the CAP is oriented towards farmers rather than consumers and agricultural exporters rather than importers. These differences make Britain an ideal candidate for a free market in agriculture with the least painful consequences for farmers. It would be much more difficult, if not impossible, for a major food exporter such as Denmark or Holland to thrive in a world of import levies and export subsidies.

But could Britain escape from the CAP without breaking up the Community? Some would say that a break-up would not matter since Britain should withdraw from the EC anyway. My own view is that EC membership is basically desirable for Britain, especially from a *political* standpoint but also as a stepping-stone to more liberalisation of world trade. Internal EC trade in industrial goods has been substantially freed and expanded despite the many remaining barriers, and the Community has negotiated within GATT successive reductions in its external tariffs. If the British economy has 'lost out' from EC membership, it has not been because of a lack of opportunities to expand our industries in a much enlarged 'home' market but because we have failed to take advantage of them. I should like those opportunities to remain.

Consequently, I do not think that we should run the risk of breaking up the Community by declaring 'UDI' from the CAP at this stage. We should let events take their course and discreetly do everything to assist the natural withering-away and/or breakdown of the CAP. The withering-away has been going on since 1968 and proceeds apace. National farm subsidies today account for more expenditure than the CAP. Many of them are in contravention of CAP rules, such as the special subsidies to French pig farmers announced early in 1984. Other national aids, such as the activities of the British Milk Marketing Boards and our protective milk hygiene regulations, are barriers to trade.

---

[1] *The Economist*, 1 November 1980, p. 54.

And the green money system has ensured that, since 1968, there has been neither truly free internal trade nor common pricing in the agricultural sector.

Most important, however, the money is running out. The Fontainebleau Summit (June 1984), which agreed to increase the EC's income by raising the limit on the Community's take from national VAT revenues, has relieved some of the pressure on the budget and thereby for reform of the CAP. But it was not a permanent settlement of the budget issue. The new income may soon be exhausted if new ways of curbing the CAP are not found and the gap between the CAP and world prices remains the same. A further increase in EC revenue would require unanimous agreement – it would be vulnerable to veto by Britain or any other member – although the summit agreement did say a second VAT increase 'may' occur after 1988. The whole issue may have to be re-opened before then, and almost certainly will if Spain and Portugal are admitted. They are expected to enter in 1986.

Britain missed an opportunity to bring the matter of the CAP to a head at Fontainebleau. Although Mrs Thatcher had said she would link an increase in revenue to a fundamental reform of the CAP, she settled for a temporary solution to the problem of Britain's budget rebate in return for agreeing to an increase in VAT contributions. The agreed budget rebate (66 per cent of the gap between Britain's VAT payments and EC expenditure in this country) will continue only until a further increase in EC revenues is required, whenever that may be, and is therefore not the permanent solution she had sought. Fontainebleau was not a triumph for Britain. We put aside completely our demands on agriculture and gained only partial success on the budget issue. However, the Community remains in a state of flux. The opportunity to end the CAP or to escape from it may come to hand at any time. It is up to the British Government to use that opportunity while ensuring that the Community survives.

I am not alone in believing that the EC could survive without a common agricultural policy. The Community could not have been conceived or born without the CAP; nor could it have survived its first 10 to 15 years. However, circumstances are now different. Hallett[1] quotes a good authority to support this view – Professor Ralph

---

[1] G. Hallett, *op. cit.*

Dahrendorf, a former EC Commissioner and Director of the London School of Economics. In his 1979 Jean Monnet lecture, Dahrendorf said:

> 'The Council of Ministers . . . has complicated an already virtually incomprehensible agricultural policy to the point at which this is little more than an instrument for Ministers of Agriculture to get for their farmers in Brussels and in the name of Europe what they would *not* get at their national Cabinet tables'.

He then added:

> 'I am convinced that European union would not collapse if the CAP collapsed. It had its place to cushion a massive migration from agriculture to industry. It may have served to balance French and German economic interests. I have yet to see one single reason why a CAP is indispensable today in order to advance the European construction.'

Towards a Free-Market Agriculture

Assuming we were able to disentangle ourselves from the CAP, a final question would be: 'What sort of national farm policy should we have?'. The essential principles of a desirable policy can be derived from the arguments already put forward; the administrative detail would be a matter for politicians and the civil service.

The main objective would be a freely-competitive agriculture without internal subsidies or protection from outside competition, thus permitting consumers and the economy as a whole to reap the benefits of international comparative advantage – in short, a free market in agricultural products (or, at least, the closest feasible approximation to it). It has already been argued that two important qualifications to this ideal are necessary: the requirement for food security and provisions against 'dumping'. Both would have to be incorporated into a new national policy.

The withdrawal of price support would have to be gradual, for the reasons mentioned above. It should be made quite clear to the farming community that price support would be phased out over, say, seven years so that farmers could plan ahead – preferably with the agreement of all the political parties, which should be possible since none need suffer any major loss (except, perhaps, the Liberals). During the withdrawal period, guaranteed prices – which could be maintained by deficiency payments of which Britain has plenty of administrative experience – could be reduced by seven steps to world market levels. If the gap was, say, 30 per cent at the beginning of the period, the price reduction in the first year would be 4·3 per cent. If world market prices rose during the period, the withdrawal would take less than seven years. If they fell, the percentage reductions would have to be increased in the later years.

Compensation scheme

For this policy to have any chance of being accepted by either farmers or politicians, it would have to be accompanied by a compensation scheme for farmers who would suffer hardship – particularly those who were heavily mortgaged or in debt to the banks – and for those who would want to leave farming. The principle of compensation is well-established for buying out the restrictive practices of trade unions and for relieving the pain of decline of old industries. Dockers, steel-workers, and miners all have generous redundancy schemes. Farmers have for many years enjoyed statutory rights to subsidies and protection; they are thus entitled to argue that their fellow citizens should compensate them for the withdrawal of rights they had been led to believe would continue indefinitely. Moreover, the price is worth paying if it facilitates and eases the transition to a more efficient and self-reliant agricultural economy.

The details of a suitable compensation scheme require more thought and discussion, although some work has already been done. In two articles in *The Guardian* 24 years ago,[1] Professor Eric Nash suggested the key principles of such a scheme:

> 'The essential requirements are that the payments should not be related to future output and should not in any way influence future productive decisions; . . . that the right to participate in them should be limited to those now engaged in farming and the payments should have a fixed term'.

He thought that the compensation payment might 'take the form of a personal annuity payable to all those at present engaged in farming for the remainder of their lives'; or, in certain instances, the capital value of the annuity could be paid in a lump sum to enable them to start a new business or to equip a larger farm. Whilst agreeing with Professor Nash's key principles, it is doubtful if it would be necessary

---
[1] Reprinted in E. F. Nash, *op. cit.*

or desirable to give equal compensation to every farmer. There seems no good reason to compensate at all farmers who have done well out of the subsidy system, or who do not have heavy borrowings, or who would still have considerable net assets, or who could still make a good income at unsupported prices, or for whom farming is an activity subsidiary to their main source of income. If people wish to farm as a hobby or as a means of earning 'pin money', there is no need to compensate them if farming becomes less lucrative. Institutional landowners which farm on their own account and have entered farming, quite legitimately, as a speculative exercise and in full knowledge of the risks of investing to take advantage of changeable government policies should also be left to take care of themselves. Like other tenants, tenants of institutions would be eligible for compensation.

If exceptions to compensation were devised along the above lines, according to income and net assets, the bulk of full-time farmers would probably still remain eligible for compensation – say 120,000 to 150,000 of them. If the government decided to make, say, £1,500 million a year available for deficiency payments (tapering to nothing after seven years) *and* compensation (increasing yearly for seven years) *together* – a sum roughly equivalent to current government expenditure on agriculture or current farming income – that would produce a generous £10,000 a year on average for every recipient. The form of compensation – which might be a basic flat-rate sum plus an amount per acre (varying with the grade of land) or according to the size of business, and which might be either an annuity for life or a lump sum (or a combination of the two) – would be for negotiation with the farmers' representatives within the total amount announced in advance.

Though critics of such a scheme might argue that it would be too costly, it would be less costly than the present or any other system of support. The rest of us would know that the end of the burden of farm support was in sight; and during the phasing-out period we would be saving the whole of the current costs to consumers imposed by the CAP and the marketing boards. In addition, there would be a gradual decline in the economic costs of all the distortions support causes. Taxpayers would benefit immediately from a further saving: that on the administration of support through the Ministry of Agriculture, Fisheries and Food, which would no longer be required. The need for a special ministry for agriculture in peace-time has always been dubious, particularly since it has tended to serve the special interests of producers to the neglect of consumers of food. The issue of its abolition should not be fudged because, so long as it remained in existence, it would constitute a threat to the free-market policy. It would fight to slow down or reverse the new policy and would remain a potential Trojan horse. Its indispensable functions could be transferred to other ministries as soon as the new policy was introduced. Health, hygiene and the veterinary services could go to the Department of Health and Social Security. The administration of the compensation scheme could go to the Department of Employment. The anti-dumping levy and the collection of statistics could be transferred to the Department of Trade and Industry. Special subsidies for the marginal farming areas and countryside policy generally could go to the Department of the Environment. Food security could be monitored jointly by the Departments of Health and Trade.

Obviously, after phasing out support the net saving in manpower and other expenditure would not equal the total current programme of MAFF (£2,087 million in 1983-84).[1] However, the 623 staff and £1,262 million expenditure[2] of the Intervention Board for Agricultural Produce which now administers CAP price support in Britain *would* be a net saving. Of the remaining expenditure and staff (11,493),[3] perhaps 80 per cent could be dispensed with.

The Pattern of an Unsupported Agriculture

It is impossible to predict accurately what the pattern of a largely unsupported agriculture would be, although we can be confident that it would be more economically beneficial to the rest of the population. Richard Body's assertion that agriculture would become more labour-intensive and that there would be a 're-invigoration of the countryside by the infusion of more farmers' seems highly improbable. It is unlikely they would be full-time farmers because we would probably move to a more extensive farming system, producing less and run by fewer professional operators. Many full-time farmers would probably become part-time, using their farms increasingly for tourism and other purposes and finding other occupations where available. The enlargement of holdings would continue as farmers left the industry altogether or sold part of their holdings (while retaining a smaller acreage for part-time use). Thus a trend already noticeable would be accelerated.

---

[1] Figures from Cmnd. 9143, *op. cit.*, p. 20.   [2] *Ibid.*   [3] *Ibid.*, p. 24.

Body's further contention that we would see a major switch to livestock away from arable farming is also open to question, although he is certainly correct in thinking that our geography gives us the biggest comparative advantage in grassland and therefore in grazing livestock production. The fact is that, in the present state of the art, it is much easier to grow arable crops consistently successfully with a high degree of certainty about output and profitability even on poor land than it is to grow and utilise grass successfully in order to produce milk and meat. Intensive livestock production (pigs and poultry and, increasingly, beef and sheep) *could* increase since it does not require the management of grass so much as business acumen, skilful husbandry, a lot of capital, and cheap feedingstuffs. A free market would provide access to the cheapest feedingstuffs, but the other requirements would remain unchanged. Livestock production based on imported feed rather than domestic grass might flourish most.

These, however, are all matters for speculation. Gazing into a crystal ball, I should like to make some final prophecies. The adjustment to a freer market, and existence in it, would not be as traumatic for farmers or their suppliers as they fear. The decreases in output caused by price reductions would be partially offset by rising productivity. Farmers would cut their costs by more co-operation in purchasing supplies, in sharing large items of machinery, and possibly also by increased use of contractors. They would become much more interested in marketing and more involved in contract selling,[1] the use of futures markets, and co-operative food processing (as the Danes and Dutch have long done), as well as in direct retailing. And if at the termination of the compensation scheme farmers still lacked the confidence to face the market-place and felt they needed some temporary or permanent minimum-income insurance, they could adopt an income deficiency-payment scheme along the lines suggested years ago by the American economist Boris Swerling.[2] The purpose of the

---

[1] The lack of incentive to improve marketing caused by the support system was illustrated in August 1984 when Bernard Matthews, the 'turkey king' of Norfolk (whose personal success story shows how a skilled entrepreneur can prosper in an unsupported sector of farming), concluded a deal with the New Zealand Meat Board to supply him with lamb worth a potential £75 million a year and involving 2·4 million carcases because no British supplier could be found. (Reported in *Farmers Weekly*, 24 August 1984.)

[2] Boris C. Swerling, 'Income Protection for Farmers – a Possible Approach', *Journal of Political Economy*, Vol. 67, 1959. The scheme is also described in P. G. James, *op. cit.*, pp. 351-355.

scheme would be to cushion farmers from severe temporary declines in income, not to raise incomes permanently. The benefits would not exceed a modest maximum and, in assessing individual incomes, account would be taken of earnings from non-farm sources. For such a scheme to be self-financing at a reasonable premium, membership would probably have to be compulsory for all farmers if a majority voted to establish it. It would not have to be sponsored or organised by government; the administration could be done by the NFU through its own insurance company.

Whenever the liberalisation of any part of the economy is proposed, those with a vested interest in subsidies, restrictions and protection try to scare politicians and the public with prophecies of doom if their particular market is freed. We saw that when proposals were put forward to abolish resale price maintenance, fixed exchange rates and exchange controls. The abolition of all three has come to pass with benefits to the consumer and the economy generally, and with far less serious consequences for those who perceived they would be hurt than they had predicted. Among others, solicitors, barristers, opticians, some shopkeepers (over freedom of opening hours), and, above all, farmers now oppose proposals to end their particular monopolies, restrictions, and protective devices. We should not pay attention to them; they have no more to fear from competition than anyone else.

Further Reading

Barker, J. W., *Agricultural Marketing*, OUP, Oxford, 1981.
Body, R., *Agriculture: the Triumph and the Shame*, Temple Smith, London, 1982.
Bowers, J. K., and Cheshire, P., *Agriculture, the Countryside and Land Use*, Methuen, London, 1983.
CEAS/CAS, *Agriculture: the Triumph and the Shame: an Independent Assessment*, Wye College and Reading University, June 1983.
Dilnot, A. W., and Morris, C. N., 'The Distributional Effects of the Common Agricultural Policy', *Fiscal Studies*, Institute of Fiscal Studies, London, July 1982.
Hallett, G., *The Economics of Agricultural Policy*, Blackwell, Oxford, 2nd Edn., 1981.
Hallett, G., and James, P. G., *Farming for Consumers*, Hobart Paper 22, IEA, London, 1963.
Hill, B. E., and Ingersent, K. A., *An Economic Analysis of Agriculture*, Heinemann, London, 1977.
James, P. G., *Agricultural Policy in Wealthy Countries*, Angus and Robertson, London, 1971.
Josling, T. E., et al., *Burdens and Benefits of Farm Support Policies*, Trade Policy Research Centre, London, 1972.
McCrone, G., *The Economics of Subsidising Agriculture*, Allen and Unwin, London, 1962.
Morris, C. N., 'The Common Agricultural Policy', *Fiscal Studies*, IFS, London, March 1980.
Nash, E. F., *Agricultural Policy in Britain*, UWP, Cardiff, 1965.
Omega Report: Agriculture Policy, Adam Smith Institute, London, 1983.
Tracy, M., *Agriculture in Western Europe*, Granada, London, 2nd Edn., 1982.

HOBART PAPERBACKS in print

2. *Government and the Market Economy* SAMUEL BRITTAN. 1971 (£1·00)
3. *Rome or Brussels...?* W R LEWIS. 1971 (£1·00)
4. *A Tiger by the Tail* F A HAYEK, compiled by SUDHA SHENOY. 1972 (2nd ed. 1978, 2nd imp. 1983, £2·50)
5. *Bureaucracy: Servant or Master?* WILLIAM A NISKANEN; with Commentaries by LORD HOUGHTON, MAURICE KOGAN, NICHOLAS RIDLEY and IAN SENIOR. 1973 (on microfiche only: £3·50)
6. *The Cambridge Revolution: Success or Failure?* MARK BLAUG. 1974 (2nd ed. 1975, 2nd imp. 1979, £1·50)
8. *The Theory of Collective Bargaining 1930-1975* W H HUTT; with Commentaries by LORD FEATHER and SIR LEONARD NEAL. 1975 (2nd imp. 1977, £2·00)
9. *The Vote Motive: The Economic Approach to Politics* GORDON TULLOCK; with a Commentary by MORRIS PERLMAN. 1976 (2nd imp. 1978, £1·50)
10. *Not from Benevolence... Twenty years of economic dissent.* RALPH HARRIS and ARTHUR SELDON. 1977 (2nd imp. 1977, £2·00)
11. *Keynes v. the 'Keynesians'...?* T W HUTCHISON; with Commentaries by LORD KAHN and SIR AUSTIN ROBINSON. 1977 (2nd imp. 1978, £2·00)
12. *The Coming Confrontation: Will the Open Society Survive to 1989?* THE DUKE OF EDINBURGH, W H CHALONER, NORMAN GASH, F A HAYEK, JULIUS GOULD, IVOR PEARCE, MAX HARTWELL, RAYMOND FLETCHER, JO GRIMOND, NIGEL LAWSON, RALPH HARRIS, ARTHUR SELDON. 1978 (cased £5·00, paperback £2·50)
13. *Over-ruled on Welfare* RALPH HARRIS and ARTHUR SELDON. 1979 (£3·00)
14. *The Emerging Consensus...? Essays on the interplay between ideas, interests and circumstances in the first 25 years of the IEA.* GRAHAM HUTTON, T W HUTCHISON, D P O'BRIEN, J M BUCHANAN and GORDON TULLOCK, A J CULYER, DAVID COLLARD, KEN JUDGE, IAN BRADLEY, COLIN CLARK, LORD CROHAM, ERIC SHARP, NORMAN GASH; with Commentaries by RALPH HARRIS, ARTHUR SELDON and JOHN B WOOD. 1981 (cased £6·00, paperback £3·60)
15. *Agenda for Social Democracy: Essays on the prospects for new economic thinking and policy in the changing British political scene.* ARTHUR SELDON, RALPH HARRIS, GEOFFREY E WOOD, A R PREST, MICHAEL BEENSTOCK, S C LITTLECHILD, LJUBO SIRC, CHARLES K ROWLEY, A I MACBEAN. 1983 (£3·00)
16. *The Poverty of 'Development Economics'* DEEPAK LAL. 1983 (2nd imp. 1984, £3·00)
17. *Markets under the Sea? A study of the potential of private property rights in the seabed.* D R DENMAN. 1984 (£2·50)
18. *Hayek's 'Serfdom' Revisited: Essays by economists, philosophers and political scientists on 'The Road to Serfdom' after 40 years.* NORMAN BARRY, JOHN BURTON, HANNES H GISSURARSON, JOHN GRAY, JEREMY SHEARMUR, KAREN I VAUGHN; with Recollections by ARTHUR SELDON. 1984 (£3·50)
19. *Choice in Education* S. R. DENNISON. 1984 (£2·50)